The Career Guide
to the Horse Industry

The Career Guide to the Horse Industry

T. A. LANDERS

DELMAR

THOMSON LEARNING

Australia Canada Mexico Singapore Spain United Kingdom United States

DELMAR

THOMSON LEARNING

The Career Guide to the Horse Industry

T. A. Landers

Business Unit Director:
Susan L. Simpfenderfer

Executive Editor:
Marlene McHugh Pratt

Acquisitions Editor:
Zina M. Lawrence

Developmental Editor:
Andrea Edwards

Editorial Assistant:
Elizabeth Gallagher

Executive Production Manager:
Wendy A. Troeger

Production Manager:
Carolyn Miller

Executive Marketing Manager:
Donna J. Lewis

Channel Manager:
Nigar Hale

Cover Image:
© PhotoDisc

For permission to use material from this text or product, contact us by
Tel   (800) 730-2214
Fax   (800) 730-2215
www.thomsonrights.com

Library of Congress Cataloging-in-Publication Data
Landers, Theodore A.
     The career guide to the horse industry / Theodore A. Landers.
        p. cm.
     ISBN 0-7668-4849-3
     1. Horse industry—Vocational guidance. 2. Horse sports—Vocational guidance. 3. Horse industry—Vocational guidance—United States. 4. Horse sports—Vocational guidance—United States. I. Title.

SF285.25 .L36 2001
636.1'0023—dc21   2001047702

NOTICE TO THE READER

To my family:

Jeriann, Kristin, Jennifer, Meghan, and Rick

Contents

Preface

*T*he basic aim of this book is to increase your prospects for finding that ideal position within the horse industry. It is simply not enough to be a talented horse person; you must also have the ability to market your talents, and that means more than just having a good cover letter and résumé. You need to possess the correct appearance, attitude, and personality. This book will assist you in these areas, as well as provide you with the basic follow-up procedures necessary to make your equine job search campaign successful.

Now that you have *The Career Guide to the Horse Industry* in hand, you are way ahead of your competition. After studying this book, you will know far more than most applicants about the best ways to cope successfully with situations that defeat so many equine job seekers. Your job search will naturally depend on the availability of the positions and the amount of competition for each opening. In some areas of the horse industry, there is an enormous demand for qualified people, while in other areas the number of qualified people exceeds the number of available positions.

The *Career Guide to the Horse Industry* will become your most valuable tool. Whether you are attempting to land your first horse-related job or returning to the equine workplace after a long absence, this book will inform you about the many careers available with horses, careers that you may find interesting and rewarding. It is written to cover the basic skills that everyone must have to successfully conduct a job search for a horse-related position.

This book is comprehensive. It not only lays out the basic skills, but it is also filled with tips and techniques that will make a difference. The job search is a series of steps, beginning with the first step of setting a job objective, and ending with the last step of accepting a job offer. In between are several steps, including writing résumés and cover letters, answering advertisements, researching, and personal interviews. This book will help you to approach your job search with a clear understanding of what potential equine employers expect of a job applicant. Finally, it will provide you with confidence in your ability to locate a position with horses that offers you both professional and personal rewards.

As an added feature, the book is accompanied by an On-line Resource™. This resource is your link to the horse industry. The On-line Resource contains eight directories to help focus your career search. Listed below are the directories available on the On-line Resource:

Directory of Professional Equine Resources

Directory of Educational Institutions Offering Programs in Equine Studies

Directory of Veterinary Colleges and Universities

Directory of Educational Programs Offering Veterinary Technology

Directory of Racetracks Offering Pari-Mutual Flat Racing

Directory of Racetracks Offering Harness Racing

Directory of Equine Periodicals and Newsletters

Directory of Breed Associations and Registries

The On-line Resource is a restricted area that requires a username and password to gain access. You will find that information listed below. You will need these to enter the restricted area, and you must enter the username and password exactly as they appear.

username: h9r8s2e5
password: 5e2s8r9h

The Directory of Professional Equine Resources is offered in the Appendix as well as in the On-line Resource.

Acknowledgments

Since 1973 I have educated both high school and adult students enrolled in the Vocational Equine Science Program at the L. A. Wilson Technological Center at Dix Hills, New York. Recently I have begun teaching Equine Science at Long Island University located in Brookville, New York. Over the years I found myself counseling my students in job campaigns designed to secure them a rewarding career with horses. I wish to express my gratitude to all those graduates who provided me with the opportunity to help them develop their equine skills, to explore the options within the entire horse industry available to them, and in doing so to change the pattern of their lives.

In the preparation of this book, many friends and colleagues provided not only encouragement but also useful information on how to conduct a successful job campaign for anyone seeking a position within the horse industry. I wish to express gratitude to the following:

Clairborne Thoroughbred Farms of Paris, Kentucky, my first equine employer

Mr. Garrett Redmond, Equine Insurance Executive

Ms. Peggy Vandervoort, Bloodstock Agent and Equine Insurance Executive

Ms. Ruth Davis, typist and valuable coworker

Mrs. Dawn Roberto, for typing and editing the entire manuscript

Mr. Phillip X. Munisteri, Educational Administrator, for granting me my first Equine Teaching position

Linda J. Ireland, for copyediting the final draft of the manuscript

I should like to express my gratitude to the following individuals and professional organizations who graciously took the time to respond to my letters and phone calls to assist me in developing and refining the subject matter and transforming it into a finished text:

American Veterinary Medical Association, Schaumburg, Illinois

W. Budd Benner, Professional Farrier, Westbury, New York

Janet Biggs, Professional Videographer and Equine Artist, New York City, New York

Brotherhood of Working Farriers Association, LaFayette, Georgia

Christine W. Brune, Executive Director, American Horse Publications, South Daytona, Florida

Charles Craig, Program Coordinator, National Animal Control Association, Kansas City, Missouri

Ruth Davis, Professional Riding Instructor, Knoll Farm, Brentwood, New York

John Giovanni, former National Manager, Jockeys' Guild, Lexington, Kentucky

Ellen Harvey, Harness Racing Communications, Colts Neck, New Jersey

Barbara Heine, PT, HPCS, American Hippotherapy Association, Woodside, California

Hooved Animal Humane Society, Woodstock, Illinois

Captain John Kmerz, United States Park Police, Washington, DC

Charlette B. Kneeland, Director, American Riding Instructors Association, Bonita Springs, Florida

Daniel Loughran, Professional Teacher and Videographer, Miller Place, New York

Dennis Manning, Certification Chairman, American Farrier's Association, Lexington, Kentucky

Jere R. Mitchell, National Association of Animal Breeders, Columbia, Missouri

George and Lori Mittelstaedt, Professional Laboratory Technologists, Brightwaters, New York

North American Riding for the Handicapped Association, Denver, Colorado

John Pawlak, Publicity Director, United States Trotting Association, Columbus, Ohio

Nicole Powell, Education Assistant, American Association for Laboratory Animal Science, Memphis, Tennessee

Kathy Rehn, Professional Travel Agent, Stonybrook, New York

Paula Rodenas, Professional Equine Journalist, Merrick, New York

Celeste Sparks, M.Sc., Equine Nutritionist, Purina Mills Inc., St. Louis, Missouri

Dr. Shannon Willoughby, DVM, American Veterinary Chiropractic Association, Hillsdale, Illinois

The author and Delmar also wish to express their thanks to the content reviewers. Their input and expertise added greatly to this new text.

Jon Wolf, Blackhawk College, Kewanee, Illinois
Charles Casada, Jasper High School, Farmersville, Texas

About the Author

T. A. Landers has been actively involved in the horse industry for most of his life. He has extensive training and experience in the field of Equine Science. He obtained a BS degree in Animal Husbandry from California State Polytechnic College and also holds an MS degree in Arts and Liberal Studies from the State University of New York at Stony Brook. Mr. Landers has taught vocational equine science at the high school level and has also taught at the university level. As a faculty member in Equine Science at the L. A. Wilson Technological Center (Western Suffolk County Board of Cooperative Educational Services), he served as FFA advisor to the Dix Hills chapter's horse judging and equestrian team. He currently serves as coordinator of the Equine Studies program at Long Island University, C. W. Post Campus.

Mr. Landers is also self-employed as a thoroughbred trainer and as a free-lance writer. He authored the text *Professional Grooming and Care of the Racehorse.* He is an active member of the Nassau/Suffolk Horseman's Association, New York Thoroughbred Trainers' Association, Agricultural Teachers Association of New York, American Horse Safety Association, New York State Agricultural Society, the American Horse Publications Association, and the Certified Horseman's Association (CHA).

1

Establishing Goals or Objectives

A large majority of equine job applicants do not have a clear idea of exactly what they are looking for in a job and they have no specific job objective or goal in mind. No one can develop a winning attitude for you and no one can set your personal goals and objectives. You must do those things yourself.

Your objective must be specific, clear, and specialized. You must be able to visualize it in your mind, and it must become a part of you. While it should be challenging, your objective should also be believable. Your employment objective must be something you really desire and something you can develop a burning desire to attain.

Your goal or objective should be detailed and reflect exactly what you want in a horse-related position. It should have a deadline for attainment. Your goal or objective should address the following questions:

Is it specific?   Does it include income goals?

Is it brief?   Is there a deadline?

Is it achievable?   Is there a location factor?

Is it believable?   Does it include a specific type of horse operation?

If you wish to be in the running for a horse-related position, you will have to do more than just answer a few newspaper or magazine advertisements. You will have to develop a total employment campaign, a campaign that will serve as a road map leading the way to your employment objective. Any goal is meaningless unless you know what you need to do in order to achieve it.

Learning what is important to you is an essential part of understanding yourself. It also will help you to clarify your career goals and will assist you in finding the right direction at this point in your life. Without any idea of where you are heading, you are undertaking a hopeless and confusing task.

Any achievements in one's life are usually accomplished slowly over a long period of time. Be prepared to break any major or long-term goals into smaller or short-term goals. For example, if your long-term goal is to become a Veterinary Assistant, your short-term goals might include:

1. working part-time for a small animal practitioner
2. enrolling in a public or private vocational veterinary training program on a high school or college level

A part-time job at a veterinary clinic may provide you with the practical experience needed to reach your long-term goal, as well as an income to allow you to pursue your education toward that same goal. The process of selecting an acceptable horse-related job involves an analysis of your personal goals, interests, abilities, financial needs, financial goals, and available time.

Evaluating Your Interests and Abilities

With a pencil and paper in hand, prepare a list of absolutely everything concerning horses that is of interest to you or that you like to do. An analysis of your abilities can be accomplished in the same manner. Write down every skill, knowledge, and ability about horses you possess. After preparing your lists of interests and abilities, analyze them. Zero in on your strongest interest and your most marketable abilities.

Evaluating Your Financial Needs

The income potential of each type of horse-related occupation that you are considering should be carefully analyzed and compared with your financial needs and wants. In some cases, for example, if you are a horse farm employee, you can be assured of receiving a weekly or monthly paycheck of a specific amount. In other situations, for example, if you become involved with horse sales or operate your own business or service, there may be no assured income and your pay periods may be very sporadic.

Evaluating Your Available Time

The amount of time and type of time you have available to devote to your horse-related position should be considered as you study each type of opportunity. Some jobs within the horse industry call for a regular schedule of a certain number of hours per day, at the same time every day, week after week. With others there is no set schedule; you work when you want to work. In some cases, the work exists steadily year round, while other horse-related jobs are seasonal.

Other Considerations

There may be some horse-related activities that sound appealing but for which you do not have a suitable background or proper training. If you find yourself in this position, a thorough study of the job-entry requirements, necessary skills, and methods of obtaining training should be undertaken. This analysis should be coupled with a review of your long-term goals. In many cases, it may be well worth the effort to enter a training program to gain the abilities necessary to qualify for an attractive equine employment opportunity.

At this point, you need to understand the difference between a "job" and a "career." A job is what you presently do at work. A career may involve a series of jobs with a shared focus or interest.

The best way to decide on a career with horses is to approach the decision step by step. Develop a plan of action beginning with a self-analysis; determine who you are, what you like, and what exactly you want. The next step is a job analysis whereby you research the job and career areas that appeal to your interests and needs. Talk with parties involved in the job and career areas that interest you. Finally, determine your career alternatives. Decide whether you will terminate your present employment or seek a horse-related job or career in a new area.

It is extremely important to organize yourself so you will be able to arrive at quick decisions about your job or career with little procrastinating. A common mistake is being vague about one's goals. It is to your advantage to have one clear distinct goal and concentrate on attaining that goal. The ideal manner in which to conduct an employment search is to be flexible, but to keep narrowing your goals until you form realistic employment goals. List important factors of your personality and job needs, and relate them to those positions you held when you were most satisfied. Think about what job-related factors made you content and most productive. What do you desire to get out of a career or life?

After you have completed your self-analysis, choose one or more employment areas that seem to suit your personality and interests. Research those positions within the horse industry that interest you most so that your choice will be realistic. For example, it is very easy to get caught up in the romance of being an internationally acclaimed rider at the Olympics and ignore the pitfalls of low salary and advancement potential until it is too late. Look before you leap into a new career. At the end of your research, you should have several major career areas and job choices that you wish to pursue in depth.

Interviewing people in the fields that interest you is by far the best method of evaluating any position within the horse industry. During this phase of your research, concentrate on getting a "feel" for the career or the job itself. The information you obtain will save you time by giving you a true picture of your ideal position. When interviewing, ask questions such as:

What are the best and the worst aspects of the job or career?

Would you work within the horse industry if you could do it all over again?

What training, education, or special skills are needed?

Are there any career opportunities presently available in the horse industry?

What type of person would you say is best suited to work with specific horse-related jobs and careers?

At this point, let us address some specific questions about your feelings that will assist you in your decision on a career with horses:

Do I definitely want to work directly with horses?

Do I possess a genuine love for horses?

Do I perform best when I am supervising other people and seeing that things operate smoothly?

Do I prefer a secure office position with the standard benefits?

Does my talent lie in administrative rather than field work?

Do I have the ability to communicate with animals?

Am I willing to devote several years to enhancing my education to become a highly skilled professional horse person?

Am I prepared to deal with the unpleasant side of the position such as illness and death of a horse?

If required, am I prepared to perform strenuous labor with an irregular schedule?

Am I willing to relocate in order to accept a position with horses?

If your answers to these questions are *yes,* it is time for you to pursue a genuine, rewarding career with horses.

Now that you have identified what you want to accomplish, you must decide how you will do it. It is therefore necessary to develop a plan for reaching the goals you have set. This plan should consist of a series of steps you will take to move yourself along steadily until you have reached your goals. Your first step will be to decide on the kind of education you need to reach your goals.

Equine Education and Training

*B*ecause of the large number of job seekers in the horse industry, many employers are now looking for potential employees who possess a formal equine education in addition to past practical experience. This large number of potential employees means that employers today can be selective in choosing a candidate to fill an open position within their organization.

Before you attempt to apply for the job of your dreams, you might ask yourself: Do I have the necessary training for my career choice? If your answer to this question is *no* then you must determine the type of education you wish to pursue. Naturally, your available time and your present financial status will be two of the determining factors.

An important part of examining a specific job as a career choice is knowing what specific educational and training requirements are needed to get and do the job. All jobs have basic educational and training requirements. Sometimes these requirements can be obtained by taking a high school level vocational training program. Some jobs have educational requirements that include post-secondary education, which means an education after completing high school. These education requirements can be achieved by attending a two- or four-year college. A student enrolled in a college program earns a degree when a course of study has been completed.

Sometimes educational requirements are offered through apprenticeship programs. Apprenticeship programs are merely training programs that take place when you first obtain a job. The training one receives in an apprenticeship program is usually controlled by either a trade union, specific employer, or both together.

Some careers and jobs require a state license as part of the basic educational requirements for that particular line of work. In these cases, one must pass a state examination before a license is issued. A license is a certificate given by a particular state which indicates that you have successfully completed a state-approved training program. It also means that you have passed a state examination that has measured your knowledge about a certain skill. Some examples of a career or job that requires a state examination are a Veterinarian and a Racehorse Trainer.

Educational requirements for the horse industry may vary depending upon the specific career or job. In any event, basic educational requirements may be expected by one or more of the following: employer, union, school, or state. You must be aware of these expectations before you make any career decision.

Many people who are seeking a career with horses do not know how to go about obtaining the proper education and training necessary for a horse-related position. There are several options available for consideration when choosing an equine education. Your educational options will fall into two basic categories: traditional college/university education, and noncollege training opportunities.

Traditional College/University Education

Whether you are an adult or a recent high school graduate interested in pursuing a career within the horse industry, several colleges and universities offer a variety of opportunities toward a future with horses. Recent changes in the economy have caused many colleges and universities to gear their existing equine programs to appeal not only to high school graduates but also to adults seeking a new career with horses. You may be wondering whether a college education is actually worth your investment of time and money. A sound college education in equine sciences can open many doors for you and provide rewards for the rest of your life.

A college education will prepare you for an occupation or for graduate and professional study. It will provide a foundation for learning and for holding a position that requires a high degree of literacy and the ability to solve problems. It will prepare you for a specific horse-related occupation, or for graduate and professional study leading to a career in a specialized area of the horse industry. In addition, a college education will:

◆ ensure that you have a high level of literacy, which will enable you to communicate your ideas effectively in speech and in writing

◆ give you the opportunity to think, to identify problems, and to consider all possible solutions before arriving at a decision

◆ teach you where to go for information, which will provide you with a tool for learning for the rest of your life.

Choosing a college or university will be one of the most important decisions of your life. Institutions of higher education differ from each other in a variety of ways, including the size and location of the college, facilities available on campus, types of equine programs offered, how they are organized, the nature of the faculty, requirements for admissions, cost, availability of financial aid, and availability to out-of-state applicants. Selecting one or more colleges you would like to attend based on your needs and interests takes a great deal of time but is well worth the effort.

In conducting your investigation, you will need to examine several important factors about the schools you have chosen: entrance requirements, tuition, housing accommodations, location, size, majors offered, financial aid available, accreditation, faculty, and whether the school functions on a quarter or semester plan. Let us now review some of the major factors that will affect your decision.

ENTRANCE REQUIREMENTS

Most colleges and universities require that all applicants meet certain admission requirements to establish whether the applicants are capable of successfully completing the college program and will be a credit to the reputation of the college. The basic selection guidelines are:

◆ entrance examination scores

◆ recommendations of teachers, counselors, employers, and so on

◆ school records (grades, attendance, and class ranking)

◆ participation in school or community activities

◆ personal interview with college officials

TUITION

The major expenses of attending any college or university are tuition, fees, and room and board. Some colleges and universities charge higher rates to nonresidents as opposed to residents of the state in which the school is located. Room and board expenses can be eliminated by living at home and attending a nearby college or university, but additional costs for daily commuting may be substantial. Other expenses such as books, supplies, and personal expenses also must be taken into consideration.

There are many financial aid programs available to the applicant to offset costs. These programs offer various forms of aid, including grants, scholarships, part-time employment, student loans, work-study programs, and government-sponsored benefits.

ACCREDITATION

Accreditation is certification that a school or program meets the requirements of a recognized agency concerning its curriculum, facilities, and so on. Attending an accredited institution is vital for those students who wish to transfer to other colleges or universities or who plan on performing graduate work in a specific field.

LOCATION

The location of the college or university has little or no bearing on the quality of education offered by the institution. Equine college programs are offered in virtually all areas of the United States. Your personal preference in regard to prevailing climate conditions may influence your decision. You should obtain some general knowledge of the community and the general area where the institution is located.

SIZE

As with location, the size of the college or university has little to do with the quality of education offered. A large college or university will offer a variety of courses and have many departments. Smaller institutions located in small towns will offer a pleasant, less hectic environment.

FACULTY

The teaching staff and professors should possess a considerable amount of experience in their area of instruction. Obtain information about their equine background and their general horse experience prior to teaching. Good teachers have a great deal of hands-on experience with horses while teaching as well as excellent academic backgrounds in their areas of instruction. Finally, consider the ratio of students per teacher before making your final college selection.

In researching equine college programs, obtain a descriptive catalog from each college and university that meets your needs. The catalog will contain information about the courses offered and other factors. If at all possible, talk to faculty and graduates, as well as students presently enrolled at the institution of your choice.

Visiting the various campuses may aid you in coming to a final decision in the selection process. This is particularly important if you have been accepted by more than one college. Walk throughout the campus and visit the horse facilities. Note whether the horse unit is located on campus or somewhere nearby and whether transportation is required to attend classes. When visiting the college horse unit, consider the following:

Is the barn neat and orderly?

Are the horses in good physical condition?

What type of stall is available (box or straight standing stalls)?

What is the type and amount of bedding in each stall?

Do the horses have adequate amounts of water available?

What is the type and quality of hay and grain being fed to the horses?

Is there a proper amount of aisle space available between stalls?

Is the lighting adequate in the barn and teaching area?

If you do not own your own horse, what type of school horses are available?

Is there adequate supervision in the barns?

Are the riding rings and indoor arena of sufficient size to accommodate a large number of students?

Are the riding areas maintained properly (watering and harrowing)?

How is the footing in the riding areas? Are the areas free of rocks, and do they have proper drainage?

What is the condition of fences?

Are there any turnout paddocks or pastures available?

Is any security present at the horse unit as well as on the campus?

Is a veterinarian and farrier readily available?

Are there any stallions standing at the horse unit? Is breeding a part of the operation?

Is there adequate parking for vehicles as well as for horse trailers and vans?

By physically examining the equine facility and answering these questions, you will be in a better position to adequately determine the degree of professionalism and quality of education being offered by a particular college or university.

In addition to visiting the horse unit, your tour of the campus should also focus on the following areas: classrooms, laboratories, library, dormitories, campus security, locations of athletic programs and extracurricular activities, public transportation, laundry facilities, restaurants, places of worship, and social and recreational facilities. By visiting these sites, you will be able to determine if the environment for learning at the college or university meets your standards.

During your visit, determine whether the college has a career development office with an active placement service for graduates, and find out about the success of recent graduates in finding productive employment or obtaining acceptance into graduate or professional schools. Be sure to inquire about the success of any alumni in your specific area of study. Also ask about the types of personal and academic counseling offered by the college. Finally, many colleges and universities, in recognition of the fact that a large number of applicants are now seeking a career within the multibillion dollar horse industry, are providing business courses as part of their equine studies programs to better prepare their graduates for a highly diverse job market within the industry. Inquire about the availability of such courses.

If you seriously desire a higher education in equine studies and you are realistic about your weaknesses and strengths, there is a equine program for you.

Noncollege Training Opportunities

VOCATIONAL SCHOOLS

College is not the only means to a rewarding career with horses, and a formal education is not limited to book knowledge alone. Many people find that the best course for them after high school is to train for skilled work at a vocational school. A vocational and technical education may be the answer to the needs of many people interested in a horse career, as well as the needs of the horse industry in general.

Vocational training with horses may be offered as a one- or two-year program while you are still attending high school. This type of program allows high school students to attend a half-day vocational horse science program during their junior or senior year of high school. These students attend their regular high school for their required academic subjects for the other half of the school day.

There is a growing number of vocational and trade schools offering an equine study program located throughout the United States. These programs may specialize in areas of training for various careers within the horse industry. Some of these schools are private enterprises; others are operated by public school systems. Applicants generally must meet a minimum age requirement, and they may need to take aptitude and general interest tests prior to enrollment. Graduates of these equine vocational training programs, which may last from several weeks to two years, are awarded a certificate or a specialized degree. In some cases, students graduating from approved programs may transfer to a four-year college program and receive full or partial credit for equine courses completed at the vocational training school.

ON-THE-JOB TRAINING

Many employers provide employees with on-the-job training to qualify them for a specific job. This training may consist of working with experienced employees until the skills are mastered. The length of time needed to complete on-the-job training programs ranges from a few days to several years. Employees working with horses may be taught only one or two phases of the required work and usually are not given the opportunity to master all the skills required for the horse-related occupation.

Trainees generally start at the lowest rate of pay. Their job security and chances for advancement are affected by their qualifications and interest level, competition from other workers, and the employment and promotion policies of the employers. Some workers remain at the same entry-level position for years, while others are given opportunities to increase their horse-related job skills and advance to more responsible duties.

On-the-job training allows you to work and learn at the same time. You may have to sacrifice salary if you lack experience. Most people learn by doing, and on-the-job training is probably the most popular method of gaining knowledge about horses. Obtaining on-the-job training may be a very informal, individualized process in which a person approaches a potential employer and an agreement is reached between the two to develop a training program for the individual worker.

WORKING STUDENT

At this point, the concept of a working student should be mentioned. A working student usually works on a horse farm or stable for a small salary that may or may not include room and board in exchange for an education with horses. For example, a working student may receive riding lessons in a particular discipline of riding such as dressage in exchange for grooming, mucking stalls, and turning out and feeding the horses that are owned by the farm or stable owner.

The working student may work from 8 to 12 hours per day for 5 to 6 days per week. In recent years, there have been numerous legal suits concerning the working student and the horse farm employer, over the concept of the exchange of an equine education for physical labor and how this arrangement complies with existing worker's compensation laws. It is very important that before accepting a working student arrangement, you investigate the basic labor laws pertaining to such an arrangement with the local Department of Labor, especially if you are a minor.

SELF-STUDY

Many job seekers improve their skills and knowledge of horses by reading textbooks and viewing how-to videos purchased from bookstores, video stores, and mail-order catalogs specializing in horses. Some public libraries offer a technical and reference section staffed with special librarians to assist you in finding such resources on specific subjects. Books and videos should deal with a specific subject, not several topics, as resources covering multiple topics tend to be confusing and superficial. Authors and instructors vary; some may be outstanding professional horse persons but lack the organization and communication skills to motivate their audience. A large number of textbooks and videos explain how to perform certain horse-related skills. There is absolutely nothing irregular about learning in this manner, but self-study is by no means a replacement for good practical experience.

CLINICS AND SEMINARS

A clinic usually allows those in attendance to physically participate in the subject matter, while at a seminar, attendees usually sit in an audience and listen to a guest speaker. These educational mediums are excellent sources for horse-related information. A clinic or seminar is usually represented by a professional in a particular field of the horse industry, such as a Trainer, Farrier, Veterinarian, Rider, Nutritionist, and so on. A local club or horse farm may sponsor a clinic or seminar. In addition, local high schools and colleges may include clinics and seminars as part of their continuing education programs offered to the general public.

VOLUNTEER

Some horse careers require a high degree of training and education; others require virtually little or no previous experience in the field. Keep such requirements in mind when considering a career with horses. It is helpful to volunteer or work on a part-time basis with horses before making a final decision on a career choice in any capacity.

Volunteers come in many varieties. They include senior citizens and young members of pony clubs, 4-H, FFA, and scout troops. As a volunteer, you will not be required to have any specific horse skills, but you will receive the necessary training after you are accepted as a volunteer. Your hours will be pre-arranged to your satisfaction. If you do volunteer, request a written evaluation of your performance from your superiors. Volunteering is an excellent means to aid you in determining your suitability for a career with horses.

PART-TIME EMPLOYMENT

During summer vacation periods, you can obtain a full-time position at a farm or stable as another method of determining your suitability for a career with horses. While attending school, whether it be high school or college, you can find part-time employment before or after school hours in order to gain some practical experience with horses.

CORRESPONDENCE SCHOOLS

A correspondence school is an excellent source of an equine education; however, it lacks that all-important factor of practical hands-on experience. On the other hand, if you have access to a horse with which to apply the skills you acquire through a correspondence school, you will definitely increase your horsemanship skills.

ON-LINE COURSES

This type of equine education provides various courses through the Internet. On-line courses may be a part of certificate or degree programs offered by colleges and vocational training schools. Most courses are offered quarterly, but some are available on a monthly basis as well. On-line courses may feature audio lecturers accompanied by on-screen slide presentations. The student may be required to complete a course evaluation and pass a posttest. In some cases, the student is able to print a certificate of completion after passing the course.

Choosing the Type of Education

In conclusion, novice students should consider the source of education that best meets their individual needs. They must take into account how much time and money they are willing to invest in obtaining their equine education. An equine education is not limited to just young people but is available to all age groups. One simply never stops learning.

3

Preparing a Résumé

Just because you are seeking a position within the horse industry, this does not mean that you should abandon the basic job-seeking procedure. When applying for any horse-related occupation, a résumé is essential. In most cases a résumé will *not* get you a job, but if properly executed, it will get you an interview. In the job search, the single most important tool is a résumé that is carefully thought out, attractively designed, and well-written. Whether or not an interview is granted results from the impression made by a résumé. The first contact with a prospective employer, and therefore the first impression made on that person, is through a résumé.

Résumé, pronounced REHZ-uh-may, is from a French word meaning "summary." A résumé is not only a summary of your experience and education, but it is also an advertisement in which you sell yourself. Like any advertisement, it should be attractive, well-organized, and capable of creating interest in its product—*you!* A good résumé is an indication to an employer that you have the proper qualifications for the job, and it should grab the potential employer's interest. A good résumé should spotlight your achievements to the reader. It should be easy to read quickly, usually within 30 to 60 seconds. Be sure to keep the résumé brief, and stress your achievements. A good résumé will provide you with an edge over other individuals seeking the same positions by

◆ allowing you to give the employer information about yourself that is not included on a job applications

◆ stating that you are serious about obtaining a particular position and that you are a true professional

◆ dictating what will be discussed during the interview

◆ highlighting your overall abilities and skills

◆ serving as a reminder to the employer of your abilities and skills after your interview is over

Why should you use a résumé? Most employers require that all candidates for a specific position submit a résumé. If you do not have a résumé that describes your horse skills convincingly, you may be eliminating yourself from being considered for positions within the horse industry that you are fully qualified to fulfill. It is important that your résumé be an accurate reflection of your skills, stating all you have to offer to an employer. One of the strongest marketing tools you can use is a good résumé.

Your résumé must be prepared before you begin your job or career search. Your résumé should be up-to-date and accurately reflect those horse-related skills you would utilize in the position you are pursuing. If you already have a résumé, your employment history may need to be updated. You want to include all of your experience and skills.

Your résumé should highlight your accomplishments as opposed to detailing your duties and responsibilities. A potential employer merely has a passing interest in your performance in a previous position. Titles and duties say little or nothing about your performance and what you have to offer a new employer that will merit you consideration. Statements of accomplishments should always expand upon your basic responsibilities and duties. Remember, it takes only 60 seconds for a potential employer to scan your résumé and decide whether to grant you an interview.

How to Write a Résumé for the Horse Industry

The following guide gives you some suggestions on preparing a résumé from scratch. It tells you what you must do before you start to write, provides you with tips on content and format, and will help you to keep important points in mind.

PRELIMINARY ACTIVITIES

It is important to analyze your background and experience as they relate to horses. The best way to do this is to prepare an abilities list that includes the following headings: Work History, Education, Personal Characteristics, and Resources. List all your abilities in order to select those that are particularly important to potential employers. This list will then provide you with the raw material you need to develop your basic résumé, and it will also provide you with the materials needed to adapt your basic résumé to the various jobs and careers within the horse industry. This listing procedure will help you to evaluate more objectively your abilities, skills, interests, and experience in relation to the type of horse-related position you are seeking. Put yourself in the place of the employer and include everything you feel he or she might be interested in knowing about you as a prospective employee.

Work History

List all your employment, including full-time, part-time, and vacation jobs as well as freelance and volunteer work. Ask yourself the following questions about each job:

◆ What was my job title?

◆ How long did I work on the job?

◆ What did I like about the job?

◆ What were the details of my job duties?

◆ What special skills or talents did I develop on the job?

◆ What references can I obtain if necessary?

◆ What did I dislike about the job?

◆ Why did I leave the job?

◆ What personality factors (imagination, initiative, leadership, and so on) aided me in being successful in the job?

Education

For the education section of your résumé, consider the following information:

◆ schools attended (omit high school if you attended college)

◆ courses taken and degrees earned

◆ scholarship honors

◆ subjects you liked best and why

◆ subjects you liked least and why

- subjects in which you excelled
- extracurricular activities such as clubs and athletics
- special skills, for example, computer keyboard, word processing, and so on

Personal Characteristics

Evaluate your personal characteristics as to their selling points and job importance. Be as objective as you can. Weigh your strong points as well as your weak points. Decide which are pertinent to the particular job with horses you desire. An honest appraisal may even help you to determine where your interests lie.

- *Identifying information*—Name, address
- *Speech*—Vocabulary, diction, voice quality, pronunciation, grammar
- *Social behavior and attitudes*—Aggressiveness, cooperativeness, temperament, adaptability, cheerfulness, patience

Note: Education and personal characteristics are important to the applicant with little or no work experience.

Resources

List all of the possible resources, leads, sources of information, contacts, and aids you may wish to use in planning your job campaign, such as:

- business associates
- organizations that may have the type of job you want
- schoolmates and instructors
- personal friends and acquaintances
- trade directories
- employment agencies—private, public, and school

PLANNING YOUR RÉSUMÉ

Analysis of Pertinent Information

Now that you have assembled all the raw material, you must process it so that you can produce an effective finished product, your basic résumé. To assist you in processing the information, answer the following questions:

- What kind of job with horses am I seeking?
- What kind of an employer will have the type of job I am seeking?
- Which part of my experience and training relates to the job I am seeking? Which part, if any, shall I minimize?

◆ Which parts of my self-analysis should be included in my final résumé?

◆ How should I arrange the parts in my résumé? Should I place education before work history?

◆ What part of my experience and background should be emphasized?

◆ Should I include military experience and membership in professional organizations?

Organization of Information

Your résumé can be organized in the following ways:

◆ *By function*—Begin with the most important function you can perform and continue with others in the order of their importance. Keep in mind what the prospective employer is looking for in an applicant. Describe each function or skill by utilizing specific examples from your experience.

◆ *By job*—Begin with your most recent job and then go backward in time. Give the name of the employer and the type of work performed.

Selection of Information

Select information that will best present your qualifications for a specific job. You must convince the potential employer that you can make a significant contribution to the organization. Always include information from your background on the degree of responsibility, the level of difficulty, and the results you obtained.

Careful selection of information, skillfully arranged, may be the determining factor in securing an interview. The résumé should reflect your abilities and qualifications. If your experience is such that you can use the same résumé without changing it for any employer, you can have your résumé professionally reproduced. You can then send it with an individualized cover letter. If you plan to apply for more than one type of job within the horse industry, however, you should prepare a résumé styled specifically for each job objective. For example, you may have a résumé aimed at a position as a Farm Manager and another aimed at a position as a Riding Instructor.

Résumé Worksheet

In order to assist you in preparing a basic résumé, it would be to your advantage to practice preparing your résumé on a résumé worksheet (see Figure 3.1) that you save on a disk. You résumé should include the following information

HEADING:

_____

First name, middle initial, last name

_____

Address, city, state, zip

_____

Area code, telephone number

OBJECTIVE:

_____

State the position you are seeking

EDUCATION: _____

Year completed, award, school, city and state

_____

List courses that will appeal to an employer

FIGURE 3.1
Résumé worksheet

SKILLS AND ABILITIES:

_____

State the first skill needed for your occupation

_____

Provide an example of how it was used

_____

State the second skill needed for your occupation

_____

Provide an example of how it was used

WORK HISTORY:

1. _____

Year left, title, name of employer, city and state

_____

Your best accomplishment

_____

List another accomplishment

2. _____

Year left, title, name of employer, city and state

_____

Your best accomplishment

_____

List another accomplishment

FIGURE 3.1
Résumé worksheet (continued)

MILITARY HISTORY:

_____

Year discharged, rank, branch of service

_____

Special abilities and talents

PERSONAL AND PROFESSIONAL DATA:

_____

Professional licenses

_____

Hobbies, interests, and activities

REFERENCES: Available upon request.

FIGURE 3.1
Résumé worksheet (continued)

Heading. This tells the employer how you can be contacted. The following information should always precede all other information:

full name

complete mailing address and zip code

area code and telephone number

Note: This information should appear in the upper right-hand corner instead of the middle of the page. The reason for this is that when a potential employer goes through a stack of résumés, your name will be easily noticed.

Occupational Interest and Goal (Objective). Specify the type of job or field of work you desire. If your background qualifies you for several jobs, list them in order of preference. If you have no experience, merely state that you are: "Seeking an entry level position in. . . ." If you are seeking your first job with horses, you are a recent graduate, or you possess a limited amount of experience, it would be best to use a "career objective." This is a statement telling your target audience exactly what type of position you are seeking.

Education. Begin with the most recent school or training program. Indicate the completion date, the degree or certificate awarded, the name of the school, and its address. Include any on-the-job training, seminars, self-study, and so on. List any courses under each school that may interest a potential employer. Information to include in this section includes:

◆ graduate schools—major and degree

◆ college—major and degree

◆ high school (do not include if you have a higher degree)

◆ specialized training

◆ professional licenses or certificates

◆ course related to your job or career choice (provide a detailed list if you have limited work experience)

◆ apprentiship programs

◆ extracurricular activities

◆ scholarships, honors, and awards

If you are a student or recent college graduate, expand upon your education section by adding significant information such as dean's list, college activities, and awards. If you were enrolled in a horse–related work–study or "co-op" program, be sure to list it in this section, but also describe the job in the work experience section.

Skills and Abilities. List the skills needed for the position you are seeking. Under each separate skill, list the various ways in which you have applied each particular skill. Be specific, adding details, accomplishments, or explanations. When you explain each skill, begin each sentence with an action verb. These action verbs will make your statements more appealing to the potential employer. The following is a list of the action verbs that are most commonly used:

achieved	*designed*	*invented*	*researched*
adapted	*developed*	*investigated*	*saved*
added	*directed*	*learned*	*served*
advised	*eliminated*	*led*	*simplified*
analyzed	*employed*	*maintained*	*sold*
assisted	*equipped*	*managed*	*streamlined*
broadened	*established*	*negotiated*	*strengthened*
built	*evaluated*	*operated*	*supervised*
completed	*expanded*	*organized*	*taught*
consolidated	*experienced*	*performed*	*trained*
controlled	*generated*	*persuaded*	*transformed*
convinced	*guided*	*planned*	*treated*
coordinated	*handled*	*processed*	*utilized*
counseled	*identified*	*produced*	*verified*
created	*improved*	*purchased*	*worked*
decided	*increased*	*reduced*	*wrote*
delivered	*initiated*	*repaired*	

Work Experience. If you have no employment history, move on to the next section. Whether your employment history includes full-time, part-time, summer, volunteer, or charitable work, do not hesitate to use it in your résumé.

Begin with your most recent employer, indicating the dates that you began and terminated that position. (If you are still employed there, use "present" for the termination date.) State your job title and the name and address of the organization. Finally, list your greatest accomplishments. Describe for each job:

◆ *Job duties*—Describe the tasks you performed, stressing those requiring the highest degree of skill and judgment. Be sure to note any specialization and any duties above and beyond your normal assignments. Indicate the use of special tools or equipment.

◆ *Responsibility*—State how many persons you supervised if you held an administrative post and to whom you were responsible.

◆ *Accomplishments*—List any outstanding results achieved. If possible, give facts and figures.

You may also describe each job by function. List the functions you performed in the order of their relevance to your job objective. Describe in brief the work you performed in each of the broad functional areas in which you qualify. This will result in a composite of duties, responsibilities, and accomplishments. Use action verbs and avoid personal pronouns, especially the first person singular. Emphasize your most recent employment. List your accomplishments, your achievements, and your duties. Write a short paragraph that summarizes your responsibilities under each job title, and then list three or four examples of specific achievements.

Military History. If you never served in the military, move on to the next section. If you do have military service, state your date of discharge, your highest rank, and your branch of service. Be sure to list any special abilities or talents.

Personal and Professional Data. This section of the résumé is optional by law. You do not have to state your age, sex, race, religion, marital status, and so on. You may wish to include some of your talents, special skills, and interests.

References. Never list references on your résumé; merely type "References Available upon Request." References usually fall into two categories: personal references and professional references.

In selecting personal references, utilize mature individuals who know you very well. It is important to obtain their consent and to advise them in advance what you would like them to say about you. Make sure they have a copy of your résumé. Ask your personal references to review the following questions before answering any inquiries from potential employers.

PERSONAL REFERENCE QUESTIONS

◆ How long have you known _____?
◆ How do you know _____?
◆ What is your personal opinion of _____?
◆ Is he or she easy to get along with?
◆ Is he or she usually punctual?
◆ What are some of his or her good qualities?
◆ What are some of his or her shortcomings?

Professional references, or employment-related references, are very important to the potential employer. As with your personal references, you should obtain the consent of your professional references. Make sure they have a copy of your résumé in their possession for review. It is advisable to send your professional references a list of questions to prepare them for the employer's inquiry.

PROFESSIONAL REFERENCE QUESTIONS

◆ How long have you known _____?

◆ Were you responsible for hiring _____?

◆ When was he or she hired?

◆ When did he or she leave your employment?

◆ What was the reason he or she left your employment?

◆ What was his or her salary upon leaving?

◆ What was his or her attendance record like? Was he or she absent from work often?

◆ Was he or she punctual?

◆ Briefly describe his or her duties.

◆ Was he or she cooperative? With coworkers? With superiors?

◆ Did he or she seem to enjoy the work?

◆ What are some of his or her good qualities?

◆ What are some of his or her shortcomings?

◆ Would you consider rehiring him or her?

Writing the Résumé

When you have completed the résumé worksheet, you can begin working on your basic résumé. You should expect to write at least two drafts. Have friends and relatives read and critique your résumé. If at all possible, have someone in the same position as that stated in your career objective review your résumé. Such a person will be able to advise you about which areas of your résumé you should stress.

Few employers read all the résumés they receive. Most employers glance at each résumé for a one-minute period. A résumé may be discarded if it is too cluttered, too long, or too disorganized. The résumés that get read are usually those that are one page in length, easy to read, and grab the reader's attention. Include only items that clearly relate to the position you are seeking—items such as skills and abilities that demonstrate your ability to do the job. In preparing your résumé, there are some basic *do's* and *don'ts* to consider.

DON'T

◆ Never enclose a photograph of yourself.

◆ Make no mention of salary, either past salaries or present salary requirements.

◆ Do not include reasons for leaving past positions.

- Do not use exotic colored paper; you should use only white or various shades of gray.
- Make no mention of membership activities or hobbies that are not applicable to your current job or career goals.
- Do not place personal references on your résumé. An employer is interested in references only after becoming serious about hiring you.
- Do not include names of spouse or children.

DO

- Be sure that a résumé is easy to read. Use clear concise sentences.
- Keep the length of your résumé short. Limit it to one or two pages depending upon your experience.
- Utilize simple language. Use the vocabulary of your chosen field.
- Emphasize your accomplishments and the skills you used to obtain desired results.
- Focus on the information that is relevant to your own job or career goals.
- Be neat. Use a one-inch margin on all four sides of the paper. Each section heading should be capitalized. Use standard 8 1/2 by 11 inch paper. (A poorly typed or structured résumé is a reflection of the applicant.)
- Have your résumé typed professionally and copied on good quality paper. Appearance makes a lasting impression. Make your résumé error-free. Always check for spelling and grammar errors.
- Write your own résumé. Review examples of résumés, but do not copy them word for word.
- Be honest. Do not lie. False statements and claims can be harmful when your background is checked.

Finally, you may possess some job-related problems that you must overcome through your résumé.

Age.   You may be considered too old for the job. In most cases, age discrimination is illegal. In order to overcome this potential problem, you might modify your résumé in the following ways:

- Do not mention any of your earlier jobs.
- Do not mention the number of years you have been working.
- Downplay or omit dates.
- Do not use a chronological format for your résumé.

Diversified Work Experience.   Listing many different jobs for short periods of time indicates to the potential employer that you are unstable and have little or no direction. To overcome this problem, modify your résumé in the following ways:

◆ Combine various jobs under a common heading or omit them completely, especially if you held a job for only a few months.

◆ Set up your résumé so it appears that you were constantly moving upward.

◆ Stress your accomplishments, but do not overdo it.

Periods of Unemployment. To deal with this problem, you can modify your résumé in the following ways:

◆ Never explain on your résumé why you are or were unemployed. You can deal with it during the interview.

◆ Stress your stability and consistency, as this will void your recent unemployment.

Reentering the Workforce Career Change. When you enter the workforce after a long absence, modify your résumé so it illustrates that your employment gap has not affected your skills or abilities.

◆ List any volunteer work as if it were a full-time position.

◆ Include any horse-related courses you have taken.

◆ Be specific about your horsemanship duties and accomplishments.

In preparing your résumé for a horse-related position, you may choose from different styles or approaches to writing your résumé. Every résumé should contain a summary of your work experience and education history, but there are three basic types of résumé styles from which to choose: analytical, functional, and chronological. The following section discusses each of these styles and provides a sample résumé in each of the styles.

Types of Résumés

ANALYTICAL RÉSUMÉ

This type of résumé lists in chronological sequence an analysis of each individual skill. Your work history and education are broken down into specific talents and each skill is listed separately. (See Figure 3.2.) The analytical style is very useful when you are attempting to change career goals. If your duties and qualifications are valuable in more than one field, it makes sense to emphasize those skills or talents by separating them from less important skills.

John Madison
15 Main Street
Garfield, Texas 61178
Home Phone: (612) 667-6001
Office Phone: (612) 667-1701

<u>Job Objective:</u> Application of my proven ability in the field to inaugurating or developing a riding program for a large-scale horse facility.

<u>Qualifications:</u>

Horse Sales:

> As Sales Manager for a modest horse farm, extended a small sales program into a full-fledged horse sales facility. Set up a breeding division. Developed an insignificant horse sales operation, manned by one person, into a sales division with 17 paid employees that provides the farm with 32% of its income.

Supervisiory Skills:

> On every job, have been required to supervise and delegate work. At no time have had fewer than 8 assistants in an established job.

Inventory control:

> Have full knowledge of inventory procedures, including audit and running inventory. Know requisition, procurement, and supply procedures.

<u>Employers:</u>

Mission Stables, Lexington, Kentucky 11680, 1981–present

Fieldstone Farms, Glendale, Wisconsin 86121, 1978–1981

Crystal Construction Co., Stowe, Texas 85017, 1968–1978

<u>Education:</u>

B.A., Centennial College, Central, Virginia 06736, 1964–1968

Participated in "Language Abroad" program, attending University of Madrid, Spain

References Supplied Upon Request

FIGURE 3.2
Analytical résumé

FUNCTIONAL RÉSUMÉ

The functional résumé places emphasis on your abilities and qualifications. It highlights different areas of employment experience and is arranged with the most important responsibilities and duties first. Each job title is followed by a brief description of duties and expertise. (See Figure 3.3.) Dates are not usually listed in this style, but bear in mind that a résumé without dates gives an appearance of being a weak résumé. Listing all the dates allows the reader to determine that nothing has been deleted and that no periods of time have been overlooked.

CHRONOLOGICAL RÉSUMÉ

As the title indicates, this style presents information in a chronological order. It is necessary however, that the presentation be in reverse chronological order, beginning with the most recent experience and moving backward in time. (See Figure 3.4.) Dates, which are always included, can be displayed in a vertical column set apart from other information. Education is treated in the same manner as your employment history. Your most advanced degree is given first, followed, in reverse order, by all other degrees and certificates. Any academic honors would be included in this group. It is the opinion of the author that this style of résumé is most effective, and therefore it is highly recommended.

Congratulations!

You have completed the résumé preparation portion of your employment campaign, and hopefully you have enjoyed the step-by-step process of writing your résumé. This chapter was designed to make you aware of your accomplishments and their value in the job market—to make you aware of the fact that you can increase your value by being willing to contribute your horsemanship skills to potential employers in a manner that fulfills their needs as well as your own.

Having a positive attitude about wanting to contribute value to a potential employer is one of the greatest assets you can have in your résumé and in the job interviews to come. You are well on your way to getting the interviews you want.

John Madison
15 Main Street
Garfield, Texas 61178
Home Phone: (612) 667-6001
Office Phone: (612) 667-1701

<u>Job</u> <u>Objective</u>: A position managing and organizing a large-scale thoroughbred breeding farm.

<u>Work History</u>:

Horse Sales Manager:

Developed energetic horse sales program extending existing sales at Keeneland and attending three additional thoroughbred markets including the Saratoga Sales. In years with Mission Stables, horse sales have increased by 95 %. Supervised 17 stable employees. Acted as paid consultant on horse sales for Northeast Regional Horse Industry Conference.
Mission Stables, Lexington, Kentucky 11680, 1981–present

Assistant Farm Manager:

Employment with this breeding farm was primarily supervisory. Directed staff of 10 in bookkeeping and payroll. Also purchased and controlled farm supplies. Worked directly under Farm Manager.
Fieldstone Farms, Glendale, Wisconsin 86121, 1978–1981

Warehouse Supervisor:

All functions, including maintaining inventory, ordering and allocating supplies, supervising assistants in office and warehouse. Many of the assistants did not speak English and had to be instructed in Spanish.
Crystal Construction Co., Stowe, Texas 85017, 1968–1978

<u>Education</u>:

B.A., Centennial College, Central, Virginia 06736, 1964–1968

Participanted in "Language Abroad" program, attending University of Madrid, Spain

References Supplied Upon Request

FIGURE 3.3
Functional résumé

John Madison
15 Main Street
Garfield, Texas 61178
Home Phone: (612) 667-6001
Office Phone: (612) 667-1701

Work Experience

1981–present   SALES MANAGER
                      Mission Stables
                      Lexington, Kentucky 11680

Developed energetic horse sales program that within two years resulted in horse sales becoming a significant source of Mission Stable's income. Acted as paid consultant on Horse Sales for Northeast Regional Horse Industry Conference. Supervised 17 stable employees.

1978–1981   ASSISTANT FARM MANAGER
                      Fieldstone Farms
                      Glendale, Wisconsin 86121

Worked principally supervisory in this organization. Directed a staff of 10 encompassing bookkeeping and payroll. Primary responsibility for each department was maintaining work schedule. Controlled and purchased all farm supplies. Directly responsible to the owner of farm.

1968–1978   WAREHOUSE SUPERVISOR
                      Crystal Construction Co.
                      Stowe, Texas 85017

Maintained inventory control and issued equipment and material. Directly responsible for ordering and allocating material to engineering department. Supervised 8 office and warehouse workers, some Spanish-speaking.

Education:

1964–1968   CENTENNIAL COLLEGE
                      Central, Virginia 06736

B.A. in Spanish Language and Literature; participated in the "Language Abroad" program, attending University of Madrid in Spain.

References Supplied Upon Request

FIGURE 3.4
Chronological résumé

4

Preparing Letters for the Job Search Campaign

*W*hen you have completed the task of preparing your résumé, you are ready to prepare those letters you will need to properly conduct your job search campaign. Before sending out any type of job search letter, you should research prospective employers. Your research should identify the following information:

- ◆ farm or stable name and address
- ◆ owner or manager's name
- ◆ services provided (breeding, racing, riding lessons, and so on)
- ◆ type of workers employed

The various types of job search letters covered in this chapter are as follows:

- ◆ cover (introductory or application) letter
- ◆ broadcast letter
- ◆ follow-up (thank-you) letter
- ◆ letter of acceptance
- ◆ letter of decline
- ◆ letter of resignation

Cover (Introductory or Application) Letter

The purpose of the *cover letter* is to get employers to read your résumé. The cover letter can be addressed to a specific individual or to an organization in response to an advertisement.

The cover letter, sometimes called the "introductory letter" or "letter of application," is usually two to three typed paragraphs in length. The cover letter is your opportunity to convince a potential employer to consider hiring you. State that you wish to be considered as an applicant for a specific position. Briefly point out how your education and/or experience could help you perform the job. Mention that you have enclosed your résumé and that you would like an interview at the convenience of the employer.

Through your cover letter you introduce yourself to a potential employer. You state your objective and sell yourself into an interview, and eventually into a job. Make your cover letter more than just a polite enclosure when sending your résumé to a potential employer. A well-written cover letter should serve to:

◆ direct a reader's attention to your strong points and away from your weaknesses

◆ show how well you know the employer's business and the horse industry in general

◆ spotlight the facts found in your résumé

Begin by planning your cover letter before you ever write anything. Read your résumé and select at least three points about your abilities, work experience, and achievements to incorporate in your letter. In order to derive the strongest points, ask yourself the following questions:

◆ Is the skill or quality essential to the job I am seeking?

◆ Can I use numbers or facts to support my claims?

◆ Does the example illustrate what I can contribute to a prospective employer?

All cover letters are made up of three basic parts: introduction, body, and closing.

Introduction. Explain in this portion of the letter that you are interested in the farm, stable, or organization or that you saw an advertisement. Seize attention by telling a prospective employer exactly what you want.

Body. This portion of the letter will determine whether a prospective employer will be interested in reading your résumé and granting you an interview. Do not be afraid to tell the employer of your qualifications. Use facts and numbers to back up your claims. Use action words, and be brief and specific. Show that you are qualified to handle the job by listing a few of the skills needed to do the job.

Closing. Close the letter by aiming for an interview with the reader. You may utilize the common indirect closing line, "I look forward to hearing from you in the near future." End your letter as positively as you opened it. Close with a "thank you for considering me" sentence. This is a common courtesy.

Examples of typical cover letters are shown in Figures 4.1 and 4.2.

Broadcast Letter

The purpose of the *broadcast letter* is to announce to a large audience that you are available for employment, so the broadcast letter must be made to appeal to a large number of prospective employers. One of the advantages of a broadcast letter is that you only have to create one letter. The broadcast letter will accompany your résumé and become part of a large direct mail campaign. A mass mailing campaign to hundreds of prospective employers will force you to send a form letter; but in doing so, you must avoid general statements like "To whom it may concern." Your success will be better if you can target your direct mail campaign, so you should design your letter so it appears that it was targeted for a single person as opposed to a large audience.

The broadcast letter should cover the same material as any job-seeking letter. State your job objective, spotlight your qualifications and skills, and request an interview.

Introduction. The opening paragraph of the broadcast letter is essential in making your letter appear as personalized as possible. Instead of writing about specific developments of each farm, stable, or organization, you can focus on the trends occurring within the horse industry. Focus on your job objective by incorporating your area of interest and the latest developments within the horse industry, and then state your qualifications. For example:

> *As the prices of thoroughbred yearlings increase every year, the role of the thoroughbred farm manager is becoming more and more challenging. Breeding efficiency, in particular, has become a vital function, and this is an area in which I possess a great deal of expertise.*

Body and Closing. The remainder of the broadcast letter should be the same as that of the cover (introductory or application) letter outlined in the previous section. A typical broadcast letter is presented in Figure 4.3.

James S. White
102 Timber Road
Omaha, Nebraska 87654
(111) 222-3333

January 5, 2001

Mr. Thomas Smith, Executive Director
Garrison Farms Inc.
10 Forrest Road
Old Town, Nebraska 87651

Dear Mr. Smith:

According to industry sources, Garrison Farms is planning to expand into the thoroughbred racing market. With Garrison Farm's innovative thinking and aggressive marketing, there is no question that the farm will soon dominate the domestic thoroughbred racing market. I'd like to join such a winning team.

I grew up on a farm and have worked with horses all my life. Having handled thoroughbreds for over six years, I find that I enjoy their high-spirited nature and the dynamic atmosphere the racetrack provides.

As an Assistant Trainer for a small racing stable, I was directly responsible for the daily conditioning of ten thoroughbreds. The stable's 2000 profits represent an increase of 32% over the previous year.

I'd like to produce similar results for Garrison Farms. With my ten years in the horse industry and my hands-on experience, I know that I can contribute a great deal to your racing division.

Enclosed is my resume. I look forward to hearing from you soon. Thank you for your consideration.

Sincerely,

James S. White

FIGURE 4.1
Cover letter addressed to an individual

James S. White
102 Timber Road
Omaha, Nebraska 87654
(111) 222-3333

January 5, 2001

The Old Town Tribune
Box R543
16 Shamrock Road
Old Town, Colorado 65187

RE: RIDING INSTRUCTOR JOB OPENING

As I plan to relocate in Colorado, your advertisement for a Riding Instructor caught my attention. Your ad stated that yours is a small farm, and that is precisely what I am looking for. I like dealing with people and in a previous position had 50 students a month. With that experience, I have learned to handle things quickly and pleasantly.

The varied activities in a position combining riding instruction and stable management sound very interesting. I have received formal education in Basic Equitation and am familiar with Dressage and Hunt-Seat Equitation if your plans include that.

My resume is enclosed for your consideration. I plan to be in Colorado in the near future and would like to speak with you about this or future positions with your farm.

I look forward to hearing from you soon.

Sincerely,

James S. White

FIGURE 4.2
Cover letter addressing an advertisement

James S. White
102 Timber Road
Omaha, Nebraska 87654
(111) 222-3333

January 5, 2001

Mr. Thomas Smith, Executive Director
Garrison Farms Inc.
10 Forrest Road
Old Town, Nebraska 87651

Dear Mr. Smith:

As the value of thoroughbreds has increased in the last ten years, the role of the professional groom has become a vital function, and an area in which I have special expertise.

I would like to learn about the career opportunities as a professional groom at Garrison Farms. Your farm comes highly recommended, and I am certain you will find me hard working, positive, and eager to learn.

For your review, I'm enclosing a copy of my resume which shows that I:

- am well spoken with a neat appearance.
- have a pleasant disposition and keen sense of humor.
- have an excellent memory for names, faces, and details.
- posses good horsemanship skills.
- am dependable, honest, and take initiative.

I look forward to hearing from you in the near future. Thank you for your consideration.

Sincerely,

James S. White

FIGURE 4.3
Broadcast letter

When mailing your résumé and letter, consider the following *do's* and *don'ts.*

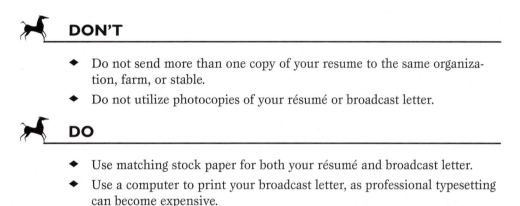

DON'T

◆ Do not send more than one copy of your resume to the same organization, farm, or stable.

◆ Do not utilize photocopies of your résumé or broadcast letter.

DO

◆ Use matching stock paper for both your résumé and broadcast letter.

◆ Use a computer to print your broadcast letter, as professional typesetting can become expensive.

Follow-Up (Thank-You) Letter

A *follow-up* or *thank-you letter* is considered a very important part of your job search. It is usually sent after an interview is over. Its purpose is to thank the interviewer, but more importantly, it is your "last chance" to convince a potential employer that you are the right person for the job. In order to make a good impression, it is best to send it on the same day as your interview. You should keep the letter short, preferably one page or less.

The effective follow-up letter should consist of the following parts:

◆ *Reminder*—Remind the interviewer that you were interviewed for the position and that you are very interested.

◆ *Job pitch*—Refer to the specific points you discussed during the interview. Build upon the impression you created during the interview.

◆ *Closing*—Thank the interviewer for his or her time and restate your interest in and qualifications for the position.

A sample follow-up (thank-you) letter is shown in Figure 4.4.

Letter of Acceptance

It is a common practice to send a letter of acceptance after an interview in which a job offer has been presented. The job offer may have been presented to you at the end of the interview, or it may have come in the form of a letter

James S. White
102 Timber Road
Omaha, Nebraska 87654
(111) 222-3333

January 5, 2001

Mr. Thomas Smith, Executive Director
Garrison Farms Inc.
10 Forrest Road
Old Town, Nebraska 87651

Dear Mr. Smith:

Thank you for meeting with me today to discuss the Director of Training position. I was especially interested in your comments about the need to expand employee education in light of Garrison Farms's rapid growth and latest expansion.

As I mentioned during our conversation, I have a proven track record in the development and implementation of thoroughbred training programs. Most recently, I was responsible for overseeing the complete care and training of ten top-quality thoroughbreds for Majestic Racing Stables of Paris, Kentucky. Proof of input was dramatic; profits increased 32 %, and the stable staff reported better understanding and higher morale.

Again, many thanks for your time and attention. I look forward to hearing from you in the near future.

Sincerely,

James S. White

FIGURE 4.4
Follow-up (Thank-you) letter

or telephone call sometime after the interview. The following points should be included in your letter of acceptance in the event your future employer has forgotten the details:

1. title and position offered
2. name of your new employer
3. date on which you will start your employment
4. agreed salary (annually, monthly, or weekly)
5. bonuses
6. medical benefits
7. vacation
8. other benefits (for example, car, house, and expense account)

An example of a letter of acceptance is shown in Figure 4.5.

Letter of Decline

If you find yourself in the ideal position in which you receive more than one offer and must decline one of them, it is customary to decline the offer with dignity, tact, and goodwill. Avoid being blunt or curt, as you may find yourself in contact with this same person sometime during the course of your career. The decline of the offer is presented in the second paragraph of your *letter of decline*. Make sure you praise the employer while declining the offer. An example of a letter of decline is shown in Figure 4.6.

Letter of Resignation

The *letter of resignation* is considered part of your job search in that you should not hesitate to utilize it whenever necessary. Basically, it is used to tell your present employer when you will be leaving, that you will help train anyone else in the meantime, and to convey the thought that you feel your experience under his or her direction has been most beneficial, regardless of how you really feel. Never burn your bridges behind you!

Word gets around very quickly within the horse industry, and you should try at all costs not to alienate any former employers. By utilizing a letter of resignation, you do the honorable thing. Make sure you send your letter of acceptance to your new employer before you present your letter of resignation.

An example of a Letter of Resignation is shown in Figure 4.7.

James S. White
102 Timber Road
Omaha, Nebraska 87654
(111) 222-3333

January 5, 2001

Mr. Thomas Smith, Executive Director
Garrison Farms Inc.
10 Forrest Road
Old Town, Nebraska 87651

Dear Mr. Smith:

This letter is to inform you that I have agreed to accept your offer of employment and I will be starting work as an Assistant Manager for Garrison Farms on February 1, 2001. I look forward to becoming part of the Garrison Farm team.

Also, I agree to the following terms of employment as set forth in your letter of January 12, 2001. These terms include a $32,000 annual salary, one month of vacation, and a full-coverage medical and dental program.

I'm sure I will find it a pleasure working for an organization that is recognized as a leader in the New York horse industry.

Sincerely,

James S. White

FIGURE 4.5
Letter of acceptance

James S. White
102 Timber Road
Omaha, Nebraska 87654
(111) 222-3333

January 5, 2001

Mr. Thomas Smith, Executive Director
Garrison Farms Inc.
10 Forrest Road
Old Town, Nebraska 87651

Dear Mr. Smith:

I would like to take this opportunity to thank you for your generous offer of employment, which I received on January 12, 2001. The position of Assistant Manager certainly is of interest to me and I greatly enjoyed visiting your facility on January 2nd.

However, I have received several offers from other employers as well, and after much deliberation I have decided to accept an offer presented to me by another employer.

In the years to come, I am sure that we will meet again. I thank you for the career opportunity you offered.

Sincerely,

James S. White

FIGURE 4.6
Letter of decline

James S. White
102 Timber Road
Omaha, Nebraska 87654
(111) 222-3333

January 5, 2001

Mr. Thomas Smith, Executive Director
Garrison Farms Inc.
10 Forrest Road
Old Town, Nebraska 87651

Dear Mr. Smith:

This letter is to inform you that I will be resigning my position at Garrison Farms effective January 31, 2001, for I have accepted employment with another organization. Prior to my leaving, of course, I will be more than happy to train another individual to assume my duties if you so desire.

I feel that my association with Garrison Farms has been most rewarding in my career. Thank you for the opportunity and the trust you have shown me.

Sincerely,

James S. White

FIGURE 4.7
Letter of resignation

5

The Equine Job Interview

All the education, research, and time you invest in your job search campaign is aimed toward this final encounter. No résumé or cover letter, however flattering, will get you a job; their purpose is to stimulate the employer's interest in your background and qualifications so that a personal interview can be arranged. Unless you are aware of how to conduct yourself during a job interview, your entire job search campaign to this point will be in vain.

The more you know about the interview process, the better you will be able to control the direction of the interview to your favor. Each interview is different, and you must prepare for each one separately. Anyone seeking a position with horses will in some way be involved with a personal interview. The interview is a highly predictable and controllable event.

Interviewing should be considered nothing more than packaging yourself for sale. The interview allows the employer to judge your qualifications and talents for the particular job opening, and it is your opportunity to convince the employer that you will make a real contribution to the overall operation. The interview also gives you the opportunity to evaluate the employer. It allows you to determine whether the position suits your specific career goals and whether the employer is of the caliber and stature you feel you wish to be associated with.

You will receive invitations for interviews because you were successful in marketing your résumé, writing letters, and following up on leads. The invitation to an interview will bring you a step closer to an actual job, but it is not a job offer—only an invitation to ask you some questions in person, to determine your suitability for a specific position. It is not over until the employer finalizes the employment process with a firm offer detailing your duties, responsibilities, salary, and benefits. This is the most desirable outcome of the interview process.

Although the interview is a selection process for interviewers, it is an elimination process for the interviewees. The job interview process is a different type of job search activity from those of writing letters, using the telephone, and meeting people for advice. Employers want the most for their money. Your performance during an interview should clearly convey the fact that you are the best candidate for the job.

Types of Interviews

Before we discuss how you should prepare for a successful interview for a position with horses, you need to become familiar with the various types of employment interviews you may encounter. You should be aware of these, since you may unexpectedly encounter one of the less common types of interviews during the course of your job search campaign. If you are asked to return for a second or third interview with the same organization, you may very well face a different type of interview than you did during the first interview. For various positions, especially professional or management-level positions, you will probably have more than one interview. In some instances you will go through three or four interviews. Each interview may consist of interviews with different individuals.

NETWORKING INTERVIEW

Usually overlooked by job seekers, *networking interviews* are an important step toward your ultimate goal of securing a job. Once you have formed an objective and identified what you do well and enjoy doing, talk with people who work in your area of interest. These interviews will give you valuable information about specific jobs and careers with horses, especially the skills and training employers expect you to have for various jobs. This type of interview will also supply you with important job information as well as help you to develop referral networks. The networks will consist of individuals who will serve as job contacts in your chosen field of interest. Best of all, network interviews usually lead to formal job interviews.

Conduct your network interviews with relatives, friends, acquaintances, and even strangers. Although most of these people will not have job openings in your area of interest, they do possess vital resources—information, advice, and networks—relevant to your job search. It is you who will request this type of interview in order to gain access to these resources. Your goal is not to obtain employment with these people, only to obtain useful information, advice, and contacts for more information, advice, and referrals.

SCREENING INTERVIEW

Employment interviews are usually conducted when there are actual or anticipated job openings. The first such interview is often a *screening interview*. As the term implies, the employer's is to screen people for further consideration. For example, a riding stable may have a pool of eight applicants for a riding instructor position. If the employer wanted to reduce the number of applicants to two for face-to-face interviews, he or she might call all the candidates to inquire about their present employment status, to obtain additional information on their qualifications, and to identify various signals indicating their suitability for the position. These telephone interviews would eliminate most of the applicants for the employer.

Employers utilize telephone screening interviews to screen applicants because such interviews are more efficient and cost-effective in eliminating a large number of applicants than face-to-face interviews, although screen interviews sometimes take place in face-to-face settings, such as college placement centers, employment firms, and job fairs. If you receive a telephone call from a prospective employer, assume that you are being screened for an interview. What you say and how you say it will determine whether you will be invited for a face-to-face hiring interview. It would be advisable to have everything you need to handle a telephone screening interview on hand so you are able to handle the call with confidence. The basic materials you should have on hand are: pencils, notepad, calendar, and résumé.

HIRING INTERVIEW

Many applicants think of the *hiring interview* as the "real" interview. In many ways it is. This type of interview will be conducted in greater depth than the screening interview. It will also have a greater effect on your future. If the hiring interview is conducted by someone from the area in which the position is open, you can expect the interviewer to ask many job-related questions. If the interviewer is from upper management or personnel, the questions asked will tend to be more general.

SERIES INTERVIEW

This type of interview is characterized by a meeting with several individuals within the same organization, usually one at a time. Following this *series interview,* the interviewers will meet to compare notes and make a collective hiring decision. Many candidates for college faculty positions encounter a series of interviews over a two- or three-day period.

The most important fact to remember during a series interview is that you must treat each successive interview as if it were your first. Bear in mind that for the person interviewing you, it is a first interview with you. Try to listen just as attentively and answer questions as fully and carefully with subsequent interviewers as you did with the first interviewer. As an applicant for a horse-related position, you may find yourself being interviewed by the stable foreman, farm manager, as well as the owners of a farm or stable.

PANEL INTERVIEW

In this type of interview you are interviewed by several people at the same time. *Panel interviews* are more stressful than other types of interviews because you are subjected to numerous questions simultaneously. Some typical panel questions are as follows:

◆   How are you going to train young horses if you do not have any experience in this area?

◆   Your college grades indicate you are not a top performer. Why didn't you do better?

GROUP INTERVIEW

Although not very common, *group interviews* do occur. If you find yourself being interviewed with several other applicants, you are in a group interview. This type of interview allows the employer to observe how the applicants interact with peers and other applicants. It is said to be an indicator of how well the individual will get along with coworkers. Usually a question is presented to the group, or the group is given a problem to solve.

Interview Locations

Most applicants expect only one person to conduct the hiring interview in an office setting. However, these interviews can vary in terms of both the number of people involved and the setting. Interviews for horse-related positions can take place in a variety of settings. Horse people are, for the most part, very time-conscious. As an applicant for a position with horses, be prepared to be interviewed in various farm settings.

The setting for the interview may vary with the type of position you are seeking. For example, if you were applying for a position with a horse insurance company, your interview would be conducted within the confines of a business office; on the other hand, if you were applying for a position as a groom, you might very well find yourself being interviewed in a tack room setting. Regardless of the interview setting, consider the interview as the most all-important phase of your job search. Without the interview you cannot properly sell yourself. Since you rarely get a second chance at an interview, you must be prepared.

Preparation for an Interview

Now that you are aware of the various types and settings of the interview, let us examine the steps involved in preparing for the interview.

RESEARCH

Research the history of the farm, stable, or organization. What is their main interest (for example, breeding, training, boarding, and so on)? Research their competitors, their reputation, the organization's ownership, and similar topics. Your effort in this research will indicate to the interviewer your sincere interest in the organization. Whatever you do, never enter an interview without knowing something about the organization doing the interviewing. It makes no difference whether you are applying for an upper- or an entry-level position.

As you obtain information on the organization, also try to determine where someone with your qualifications would fit in and exactly how you could personally contribute to the future growth and development of the organization. Many of the questions you will be asked will relate to your knowledge of the organization. For example, you may be asked:

◆ Why do you wish to work for our organization?

◆ What do you know about our organization?

◆ Why do you think you would be successful working for us?

All of these questions can be answered thoroughly if you have good background information on the organization and a clear understanding of your own qualifications.

In addition to researching the organization, try to conduct research on the individual or individuals who will be interviewing you. This information may be obtained from people you know within the organization and from promotional literature published by the farm, stable, or organization.

Developing a profile on each employer you plan to visit not only increases your ability to impress the interviewer but also is an important first step in reducing your "fear of the unknown." As you become adept at doing your homework for each employer, you will eliminate a large percentage of the unknown from the interview.

Your research should address several questions that will yield useful information for the interview. Your first set of questions should relate directly to the organization you are researching.

RESEARCH QUESTIONS

- Who are the key parties in this organization?
- What are the major services offered by this organization?
- How large an organization is it in terms of property, employees, and horse population?
- Where is the organization operating other than this location?
- How is the organization viewed by its clients, suppliers, and competition?

Another set of questions should focus on the interviewer and the interview situation.

INTERVIEW SITUATION QUESTIONS

- How many interviews must I endure with this organization before they make a final decision?
- Who normally conducts the interview? (Consider that person's position, gender, personality, style, and so on.)
- What type of interview does the organization usually conduct?
- Where will the interview(s) take place?
- Will I be subjected to any type of testing? If so, what type?

Employers want to know how well you will fit into the organization—in other words, how intelligent, competent, honest, and enthusiastic you are. This will be the basis of the hiring decision. The more you know about your interviewer, the better you should be able to manage the interview. After you organize all the materials you have gathered, analyze and reflect on them. You want to go into the interview ready to convince the employer that it would be in the best interest of the organization to hire you. The time and effort you spend in preparation for the interview will not go unnoticed by the employer conducting the interview.

Research Tools

Trade Publications.   There are many horse-related trade publications on the market covering all breeds and all types of equestrian topics. These publications are available in the form of journals, newsletters, magazines, and newspapers. For example, the thoroughbred racing and breeding industry relies on *The Blood Horse Magazine,* which is the official publication of the Thoroughbred Owners and Breeders Association. If you are not certain about which specific trade publications pertain to your area of interest, check the following directories; these may be available at your local library:

Standard Periodical Directory
Oxbridge Communications
150 Fifth Avenue, Suite 302
New York, New York 10011
(212) 633-2938

The Source (Thoroughbreds)
The Blood Horse Magazine
P.O. Box 4038
Lexington, Kentucky 40544
(606) 276-4450

Horse Industry Directory
American Horse Council
1700 K Street NW
Washington, DC 20006
(202) 296-4031

Computers. You may utilize your personal home computer or the CD-ROM index at your local library to aid you in your research. A computer will allow you to explore many of the sources mentioned earlier such as directories and newspapers. Performing research on a computer can save you a great deal of time but also can be very expensive. Computer expenses may include the following:

◆ monthly fee for subscribing to an Internet service
◆ applicable fees for local telephone use
◆ costs for paper and ink cartridges for printing
◆ electric utility costs for operating the computer and printer

Professional Horse Industry Associations. You should take advantage of the various professional horse industry associations when doing your research. These associations can provide you with important information through their official newsletters and their journals and magazines, which usually include help-wanted advertisements and may also contain articles pertaining to the latest trends within the horse industry. Membership lists of professional association will assist you in obtaining new contacts.

Using the Farm, Stable, or Organization as a Research Tool. Once you have established which organization you are interested in, you can utilize the organization itself as a primary source of information. Contacting the establishment where you will be applying for a position saves you a great deal of time and effort. The organization is best equipped to give you the answers to your questions about their operation. Calling or writing for a brochure is one of the best methods of obtaining information about a particular organization.

Most interviewers expect you to know something about their business when you are being interviewed. What should you know? To assist you in answering this question, the following outline will aid you in exploring a particular farm, stable or organization.

Name of farm, stable, or organization: _____

Name of owner, manager, or CEO: _____

Location of farm, stable, or organization: _____

Number of employees: _____

Nature of business: _____

Future services planned: _____

Competitors and their business: _____

Using Present and Past Employees of an Organization as a Research Tool. These people can give you the real lowdown on the organization. Do not be afraid to ask questions like:

- What type of skills and education is required to perform this job?
- What are the advantages and disadvantages of this type of work?
- How would you describe a typical workday?
- What do you dislike the most about your job?
- What is considered to be the normal salary range for this type of work?
- How did you go about securing this job?
- What type of employee are they looking for?
- What is the best way to approach the employer?
- What is the future for this type of work?
- How long before you receive a raise?
- How long before you receive a promotion?

With detailed and reliable information, you will be in a position to determine the best way to sell yourself.

What to Bring to the Interview

As soon as you are invited to a job interview, assemble all the information you need, organize it around the specific position, and prepare for your interview.

SUPPORT MATERIALS

Depending on the position, be prepared to bring your personal portfolio. Your portfolio should consist of a collection of your support materials. Include both originals as well as several copies of all support materials. What you bring to the interview will depend a great deal on the position for which you are applying. Regardless of the position, however, be prepared to bring the following to the interview:

◆ *Letters of recommendation*—If you have any letters of recommendation that attest to your past accomplishments and achievements, be sure to bring them to the interview, and do not be afraid to show them to the interviewer.

◆ *Résumé*—Bring at least two or three copies of your résumé to the interview. If you are interviewed by more than one individual, make sure each of them has a copy of your résumé.

◆ *References*—References are *not* to be included in your résumé or broadcast letter; instead, provide a list of references at the time of the interview. The references you use should be contacted before the interview so they are aware that you are using their names as references. It is recommended that you advise them of the position for which you are applying. Also, it is important not only that you are prepared to supply references, but also that your references are prepared to provide favorable information about you.

◆ *Awards, certificates, licenses, and scholarships*—Bring any awards, certificates, licenses, or scholarships you have received that pertain to the position you are seeking. These materials should be available for review by the interviewer. For example, if you are interviewing for a position on a horse farm, bring copies of your valid driver's license to indicate your ability to drive, as you may be asked to drive farm vehicles including a horse trailer. If you received an award or a certificate of achievement for a horse-related project as a member of a horse club such as 4-H, Pony Club, or FFA, by all means have copies available for the interviewer.

◆ *Transcript of grades*—Although seldom requested, a transcript of grades is of the utmost importance to applicants seeking an entry-level position with horses, especially those who are recent high school or college graduates. In this case, your work experience with horses, if any, would be limited, so your grades would be crucial to the interviewer.

Appearance for the Interview

Your overall appearance will have a direct impact on the success of your interview. *Appearance* refers to your dress, your physical appearance, and your mannerisms. The first 5 minutes of an interview are considered to be the most important, as that is when you project your image. A large part of that initial image is a direct result of your appearance.

DRESSING FOR A SUCCESSFUL INTERVIEW

Dress is considered a way to communicate in a nonverbal manner. If you are dressed in a sloppy manner, most people will consider you a sloppy person. A basic rule for dressing is to dress like the interviewer is likely to be dressed—but better. Different occupations with horses require different styles of dress, so there are no set rules on how to dress for an interview. A confident job applicant looks different from someone lacking in confidence. People have a tendency to dress in a manner that reflects their self-image. Those who project a "winning" attitude have a tendency to dress in neat, well-tailored clothing that conveys their success and confidence. You should dress as if you were going to work, only a tad better.

As part of your interview preparation, review your wardrobe and select those items that can be put together to create the image you wish to project. You should consider the following:

- Be sure your clothes fit properly.
- Dress appropriately for the position for which you are applying.
- Avoid excessive jewelry, aftershave, perfume, and so on.
- Stick to basic colors.
- Be sure clothing is clean and wrinkle-free.
- Be sure your shoes or boots are clean and polished.

Your goal is to appear confident, successful, and professional. You can accomplish this by wearing something other than a formal suit. Do not wear a suit if you are interviewing to become a groom. On the other hand, do not wear blue jeans if you expect to become a horse insurance executive.

PHYSICAL APPEARANCE

In addition to your attire, you must be aware of your physical appearance. At the time of the interview you should be awake, alert, and exhibiting enthusiasm. Be sure to arrive for the interview looking as clean and neat as possible. This includes your hair, fingernails, and teeth. Your hair should be clean and combed, and women as well as men should have a conservative hairstyle. Allow plenty of time to arrive for the interview; by doing this, you will be less tense and nervous and have more time to compose yourself before the interview. You want to convey to the interviewer through your appearance that you are healthy, alert, well-rested, and confident.

MANNERISMS

During the interview, the interviewer will be noting your mannerisms. The following are some important behavior patterns to consider:

- Make good eye contact with the interviewer, but do not overdo it.
- Sit upright, leaning slightly toward the interviewer. Do not slouch in your seat.

◆ Vary your tone of voice to convey enthusiasm and interest.

◆ Speak clearly. Do not mumble.

◆ Smile periodically throughout the interview, as this will convey a positive image.

◆ Avoid folding your arms across your chest.

◆ Use gestures to enhance your conversation.

◆ Listen carefully and focus your attention on what the interviewer is saying. Nod in agreement occasionally; but do not overdo it.

◆ Avoid interrupting the interviewer during the interview.

◆ Do not monopolize the conversation.

◆ Never chew gum or smoke during an interview.

◆ Always think carefully before speaking.

Salary and Benefits

Generally speaking, the purpose of your initial screening interview is for the interviewer to determine your qualifications for the position. You should not discuss salary and benefits at the initial interview unless the interviewer wishes to discuss the subject. However, if you are invited for a return interview, be prepared to seriously discuss your salary and benefits. After you have convinced the interviewer that you are the right person for the job, the main topic becomes money.

Salaries are for the most part negotiable. They are usually based upon the position rather than the individual. Salary should be considered by the applicant to be one of the most important aspects of the interview. Before discussing your salary, you should know the approximate salary range for the specific position you are seeking, know what you want in the way of salary, and have a figure in mind. In any negotiations involving income, emphasize the contributions you can make in return for the salary you are seeking. It is considered unprofessional and inappropriate to discuss other offers you have received during an interview.

In addition to a base salary, your offer may include other forms of compensation for the job. You must consider overtime possibilities, bonuses, salary in lieu of vacation time, vehicle use, housing benefits, commissions, and perhaps the use of a stall free of boarding fees. Benefits are considered to be equally as important as the base salary during negotiations. Benefit packages may include a retirement plan, medical plan, life insurance, and disability insurance. Bear in mind that a low salary with many benefits can prove to be worth a great deal more than a high base salary excluding such benefits.

The issue of income is usually brought up by the interviewer through questions like:

◆ What income range are you seeking?

◆ How much money are you earning now?

◆ What was your salary for the past two years?

Avoid mentioning any definite salary figures until you know for a fact that the organization wants to hire you. One method to do this when replying to the interviewer's questions about income is to reply in the following manner: "I'm currently earning in the mid-twenties, and while I don't wish to take a step backward, I am more interested in the right job opportunity, an organization where I can make a meaningful contribution." If you reply in this manner, most interviewers will realize that you are flexible in your salary demands and will continue with the interviewing process.

The key to successful salary negotiations is to think about salary thoroughly in advance of the interview. Have a salary figure in mind, and know why you are worth it and why it would be advantageous for the organization to pay it.

Typical Interview Questions

Before discussing specific questions asked by the interviewer, let us first examine the various types of questions asked:

◆ *Close-ended*—This type of question can usually be answered easily, as the employer is seeking specific information about an applicant. Example: Do you enjoy working outdoors?

◆ *Open-ended*—This type of question, which usually begins with *How, What,* or *Why,* allows the applicant to answer at great length. Example: Why do you wish to work for our organization?

◆ *Broad-based*—This type of question is designed to allow the applicant to demonstrate his or her ability to think about a broad subject area. Example: Tell me about your work experience over the last two years.

◆ *Self-analysis*—This type of question forces the applicant to reflect on his or her own personality and abilities. Example: What is it about yourself that allowed you to perform so well while attending college?

◆ *Leading*—This type of question leads the applicant to an obvious answer. Example: Would you be interested in working for a breeding farm as large as ours?

Although you cannot anticipate the exact questions the interviewer will ask, you can anticipate the most important ones. Expect questions regarding: education, career goal, work experience, personality, and relationship to others.

The purpose of any question asked during an interview is to determine if you are able to perform the tasks as stated in the job description. Following are examples of specific questions you may be asked during an interview.

INTERVIEW QUESTIONS ABOUT YOUR EDUCATION

◆ Describe your educational background.

◆ Why did you major in _____?

◆ Why did you choose _____ University?

◆ What was your grade point average in college?

◆ What subjects did you enjoy the most? The least?

◆ Did you hold any leadership positions?

◆ What were the reasons for such low grades?

◆ Do you feel you did your best while in school?

INTERVIEW QUESTIONS ABOUT YOUR PAST EMPLOYMENT

◆ Why did you change jobs so frequently?

◆ What horsemanship skills did you enjoy the most? The least?

◆ What was your reason for leaving your last employment?

◆ Can you tell me about some of your accomplishments with horses in each of your past positions?

◆ What did you like most about your superiors? The least?

◆ Were you ever dismissed from employment? If so, what were the reasons for your dismissal?

◆ Which of your past horse-related positions did you enjoy most? Why?

INTERVIEW QUESTIONS ABOUT YOUR CAREER GOALS

◆ Why do you wish to work for our organization?

◆ Why do you feel that you are qualified for this position?

◆ Ideally, what would you like to do?

◆ Give me some reasons why we should hire you.

◆ In what way do you feel you can contribute to our organization?

◆ How much do you feel you are worth for this position?

◆ When will you be able to start?

◆ How did you find out about our organization?

◆ How do you feel about relocating, traveling, and working overtime?

◆ What is the minimal salary that you will accept?

◆ How do you perceive yourself 5 years from now?

◆ What are your short-term goals and long-term goals?

◆ Are you presently considering employment with other organizations?

◆ How many people have you supervised?

◆ Have you ever prepared a budget?

◆ Would you be willing to work overtime in order to complete a job?

The purpose of listing these various interview questions is not only to allow you to become familiar with the questions but also to allow you to develop the appropriate answers. You should develop your responses and practice them until they feel natural. By doing so, you will give the interviewer the impression that you just thought of the responses during the interview. By practicing with a friend or in front of a mirror, you will improve your confidence immensely.

The interview questions listed here are by no means all of the questions you will ever encounter during an interview. Answer any questions asked by the interviewer honestly and directly, and most of all, present a positive attitude as much as possible throughout the interview.

Illegal Interview Questions

Title VII of the Civil Rights Act of 1964 deems it illegal to discriminate on the basis of race, sex, religion, and national origin in making personnel decisions. Personal questions concerning your physical makeup are also illegal under the law unless they can be shown to directly relate to occupational qualifications. You may not be asked any illegal questions, but that possibility is always present, so you should be prepared. The following questions are considered illegal:

◆ How much do you weigh?

◆ How tall are you?

◆ Do you have an arrest record?

◆ Are you a member of any social, religious, or political organization?

◆ Do you attend a house of worship on a regular basis?

◆ Have you ever been hospitalized? For what?

◆ Are you presently taking any medications?

◆ Have you ever been treated for drug abuse or alcoholism?

◆ Do you have any major disabilities?

◆ What is your age?

◆ What is your personal financial situation?

◆ Are you presently married, divorced, separated, or single?

◆ Do you own or rent your home?

◆ Do you presently live with anyone?

◆ Do you practice birth control?

◆ How does your spouse feel about you working?

◆ Do you have any children?

◆ Do you still have any children living at home?

Most illegal questions are asked out of ignorance. Be prepared to deal with the situation, should it arise during an interview. Unless a question is truly offensive to you, it is best not to overreact. Obviously the degree of offensiveness will vary with each individual job applicant. If you honestly feel that a question is seriously offensive to you, then by all means let the interviewer know that the question is inappropriate and that you object to the question.

If you have a problem with a particular question and you feel you have been a victim of discrimination, contact the federal Equal Employment Opportunity Commission (EEOC) to file a legitimate complaint. There are also state laws, known as fair employment practices laws, that deal with discrimination. File your complaint with both federal and state agencies within 180 days of the incident.

Questions to Be Asked by the Applicant

We have been focusing on those questions generally asked by the interviewer; however, there are certain questions that should be asked by the job applicant. You need to ask questions throughout the interview in order to obtain information about the position and the potential employer. You should also realize that the interviewer will be judging your confidence, intelligence, interest, and personality on the type and number of questions you ask during the interview. The following are some general questions that should be asked by the job applicant sometime during the course of the interview:

◆ What will my duties and responsibilities entail?

◆ Was this position recently created?

◆ How does this position fit within the overall operation?

◆ What type of individual are you looking for?

◆ What do you consider the best qualifications for this position?

◆ What do you expect of me?

◆ What is considered to be the normal salary range for this position?

- What is your organization's policy concerning promotions and advancement?
- Can you describe a typical day on this job?
- What is the career path for this position?
- How long has this position been vacant?
- Who will be my supervisors? Can you tell me something about these people?
- What is the largest single problem facing your staff now?
- What are the primary results you would like to see me produce?

Interviewers usually end interviews by asking, "Is there anything else you would like to add?" You should always answer *yes* to this question. If during the interview one of your qualifications was overlooked, now is the time to present your case. You may also use your response to this question to restate your qualifications for the position. If you feel that the interview has not gone well for you, it is at this point that you should counter all the negative aspects about yourself that emerged during the interview.

Remember, if you are in competition with other applicants for the same position, the employer will utilize the interview to seek out negative information on each applicant. Most applicants are rated by an interviewer according to a series of factors that summarize the ability of each applicant. Although *applicant rating forms* are not widely used within the horse industry, it is in the best interest of the applicant to become familiar with such forms. A typical applicant rating form is shown in Figure 5.1.

At the end of the interview, the interviewer will complete a rating form or compile his or her notes pertaining to the interview. Therefore, it is very important to leave a lasting good impression. End the interview in the following manner:

- Restate that you can perform the job.
- Offer to provide any additional information.
- Ask when a final decision will be made.
- Thank the interviewer.

After the interview, sending a follow-up (thank-you) letter is very important regardless of the outcome of the interview. See Chapter 4 for a sample of the follow-up (thank-you) letter. This letter will set you apart from most applicants and help to expand your friendships in the job market in addition to helping you lock up job offers.

Applicant's name _____

Position for which applicant is being considered _____

Date of interview _____

Factors (rate on a 1 to 4 scale, 1 being poor and 4 exceptional)

APPEARANCE

__ dress   __ grooming   __ manners   __ neatness   __ mannerisms

COMMUNICATION SKILLS

__ grammar   __ animation   __ listening skills   __ thoughtfulness   __ persuasiveness

INTERPERSONAL SKILLS

__ handshake   __ posture   __ eye contact   __ sensitivity   __ openness   __ humor

MOTIVATION

__ ambition   __ commitment   __ energy   __ goal-oriented

INTERVIEW PREPARATION

__ knowledge about organization   __ knowledge about position

INTERVIEW BEHAVIOR

__ confidence   __ questions asked   __ response to questions   __ punctuality
__ organization

ATTITUDE TOWARD WORK

__ dedication   __ ethics   __ enthusiasm   __ independence   __ commitment
__ discipline   __ team orientation

SKILL MATCH

__ education   __ experience   __ trainability

COMMENTS: _____

_____

_____

SUITABILITY FOR POSITION

__ Do not offer   __ Possibility   __ Offer

FIGURE 5.1

Applicant rating form

6

The Employment Application

When you are applying for a position with horses, you are sure to be asked to complete an *application form*. This request may occur before, during, or after the interview. The application form serves the following functions:

- allows the employer to screen applicants quickly
- states facts about the applicant
- provides the employer with a supply of future employees for a specific position
- sets the stage for the interview

Before you can complete an employment application form, you must be familiar with your résumé in detail. You will need to know information about your past employment, such as exact dates of employment, job titles, and so on. This information appears on your résumé; therefore, any application form you complete will support the information already presented on your résumé. Differences in dates, responsibilities, and names between the application form and résumé can raise serious doubts about you in the mind of the equine employer who is reviewing your application and can place your overall credibility in jeopardy. To aid you in organizing the data from your résumé, complete the application worksheet shown in Figure 6.1.

IDENTIFICATION

Name _____   Social Security No. _____

Present Address _____   State _____   Zip _____

How long at this address? _____   Home Phone _____

Previous Address _____   State _____   Zip _____

FAMILY INFORMATION

Marital Status _____

Spouse's Name _____   Social Security No. _____

Address _____   Occupation _____

Employer _____   No. of Years Employed _____

Children's Names and Ages _____

Parent or Closest Relative Name _____

Address _____   Phone No. _____

Occupation _____   Employer _____

HEALTH INFORMATION

In general, your health is _____

Have you ever had: Heart condition ____, Seizures ____, Tuberculosis ____, Back pain ____,

Dizziness ____, Any other condition that would affect your performance _____?

If you answered *yes* to any of the above, explain. _____

_____

How many days of work (school) have you missed in the past two years? ____ Explain. ____

_____

Have you ever received Workers' Compensation? _____

Have you ever received Unemployment Compensation? _____

FIGURE 6.1

Application worksheet

EMPLOYMENT HISTORY

Name of Employer _____

Address _____

Dates of Employment _____

Type of Organization _____

Size of Organization/approximate number of employees _____

Approximate annual sales volume or annual budget _____

Position held _____

Earnings per month/year _____

Responsibilities/duties _____

_____

_____

_____

_____

Achievements or significant contributions _____

_____

_____

_____

Demonstrated skills and abilities _____

_____

_____

_____

_____

Reason(s) for leaving _____

_____

FIGURE 6.1
Application worksheet (continued)

EDUCATIONAL DATA

High School Name _____

Address _____

Dates Attended _____

Year Graduated _____   Type of Degree _____

Student Activities _____

_____

Memberships _____

College Name _____

Address _____

Dates Attended _____

Year Graduated _____   Type of Degree _____

Major _____   Minor _____

Educational Highlights _____

_____

College Name _____

Address _____

Dates Attended _____

Year Graduated _____   Type of Degree _____

Major _____   Minor _____

Educational Highlights _____

_____

FIGURE 6.1
Application worksheet (continued)

VOLUNTEER EXPERIENCE

Place _____

Dates _____

Description of Duties _____

_____

_____

Place _____

Dates _____

Description of Duties _____

_____

_____

ADDITIONAL TRAINING OR VOCATIONAL COURSES

List any vocational courses, on-the-job training, and military or other training in this section.

Course _____ Date taken _____

Skills learned _____

_____

License or certificate earned _____

_____

Course _____ Date taken _____

Skills learned _____

_____

License or Certificate earned _____

_____

HOBBIES

List hobbies or activities that you enjoy. _____

_____

_____

FIGURE 6.1
Application worksheet (continued)

PROFESSIONAL MEMBERSHIPS

_____

_____

_____

_____

LICENSES/CERTIFICATES

_____

_____

_____

_____

Expected Salary Range: $ _____ to $ _____

Acceptable amount of on-the-job travel: _____ days per month

Areas of acceptable relocation: _____

_____

_____

Date of Availability: _____

Is it all right to contact your present employer for references and/or verification of

employment? _____

MILITARY EXPERIENCE

Service _____   Rank _____

Dates of Service _____

Reserve Status _____

FIGURE 6.1
Application worksheet (continued)

LIST TWO REFERENCES FOR EACH TYPE

Academic References:

Name: _____   Title: _____

Address: _____   Phone No.: _____

Relationship: _____

Name: _____   Title: _____

Address: _____   Phone No.: _____

Relationship: _____

Employment References:

Name: _____   Title: _____

Address: _____   Phone No.: _____

Relationship: _____

Name: _____   Title: _____

Address: _____   Phone No.: _____

Relationship: _____

Character References:

Name: _____   Title: _____

Address: _____   Phone No.: _____

Relationship: _____

Name: _____   Title: _____

Address: _____   Phone No.: _____

Relationship: _____

FIGURE 6.1

Application worksheet (continued)

Because the application form plays a significant role in your overall job search, it is important for you to thoroughly understand the questions so you can utilize the application form to reflect your true horsemanship skills and qualifications for a specific horse-related position. The following sections discuss the types of information typically covered in an application form.

Personal Information

Employers want to know who you are, where you reside, how to contact you, and similar information. The following topics are those most likely to be addressed.

Name. Be sure to write your name in the requested order (last name first, then first name, and so on). Use your middle name or initial only if indicated on the application form. Use your full legal name, not nicknames. If you are required to sign the application form, sign your name as stated on the application.

Social Security Number. Be sure you have this number correct. An error here can cause problems at a later date with your salary checks, benefits, and taxes. If you do not have a social security number, apply for one at your local social security office.

Address. In addition to being a place to send your mail, your address may indicate to an employer how long you have resided in a particular city. Employers may request more information than just your present street address, city, and zip code; they may ask how long you have resided at this address, whether you own or rent your residence, and so on. If you have lived at your present address less than a specified time, you may be asked to provide the same information for a previous address.

Telephone Number. An interested employer may attempt to reach you by telephone, so be sure you provide your correct telephone or fax number. Include the area code.

Physical Characteristics. Your height and weight may be requested, particularly if a job (such as an Exercise Rider) has certain physical and weight requirements.

Race, Religion, and National Origin. As stated in Chapter 5 on Interviews, it is illegal to select employees based on these factors. In fact, it is now rare to find these factors listed on an application form, but if you do, you may leave this section blank—although you must use good judgment in doing so.

Date of Birth. Most applications no longer ask for your date of birth or age, as it is illegal to hire based on this factor. However, you may have to be a certain age to qualify for a specific position. To drive a horse van or truck, for example, you must be a certain age in order to obtain a driver's license. You may need working papers if you are underage. Check with your state Department of Labor office to obtain the required forms and information.

Citizenship. If you are not a citizen of the United States, you may be asked to prove your ability to work legally.

Family Information

Marital Status. When you are asked this question, you may write "single" or "married." Do not include details such as divorced or separated. Unfortunately, some employers judge stability based on marital status.

Health Information

Questions concerning your health are asked to determine your dependability as an employee. Chances are you will be screened from consideration if you have a history of health-related absences.

General Health. Unless you have a major medical problem that will keep you from performing your job, your response should be "excellent." Never use the terms *average* or *good* without a good reason. These responses will usually get you eliminated.

Disabilities and Physical Limitations. Any physical limitation is considered a disability on a job only if the job requires a task that you are unable to perform.

Education and Training

Within this section, you should present all your education and training in as positive a manner as possible. Use every available space to present something positive about yourself. Be sure that dates are correct when indicating periods of education and dates of graduation. Have the addresses of schools and training facilities on hand.

Military Experience

Some applications ask for significant information, and others do not. If there is no military section on an application or if that section does not let you present your training and experience, utilize the education or work experience section of the application to indicate military training and experience.

Position Desired

Most applications ask what position you are seeking. How you respond to this question is very important.

Job Objective. If you are aware of a specific job opening with an employer, you may utilize that job title or some variation of it. For example, if you were looking for a position as a Farm Manager, you might write "management and related tasks." This would leave your options open for other jobs that might interest you. The job objective that you write on your application emphasizes the skills and experiences throughout your application.

Salary. It is best not to state a specific salary. Write "salary negotiable" or "open," as you do not wish to be screened on this factor. If you are required to write something, write in a wide salary range, such as "mid-twenties," "low to mid-thirties," or "$8.00 to $10.00 per hour."

Hours. You may be asked if you are willing to work evenings, weekends, or holidays. The best answer to this question is "will consider" or, if you do have a preference, "prefer daytime hours but will consider other shifts."

Work Experience

Employers will examine what you have done on previous jobs as an indicator of whether you can perform the job you are seeking. You must present yourself as a person who has an excellent chance of succeeding at the job you are seeking.

Employment Gaps. If you have gaps in your work history, present them in a positive manner. For example, if you went to school, received career counseling, or did anything else during the time between jobs, by all means mention it, as it provides a reason for the gap.

Job Titles. If for some reason the job title you held does not describe your responsibilities or duties, consider amending it. For example, if you were a "camp counselor" but supervised a group of people, you could use a more descriptive title such as "Department Head, Equine Services." Select new titles that would be helpful in qualifying for the new job.

Reasons for Leaving Past Jobs. Never write "fired" as a reason for leaving a previous position. Restate a discharge in a positive manner, such as "left to further my education" or "accepted a position for a higher salary."

Name of Supervisor. If you are concerned about what your ex-supervisor might say, consider giving the name of another responsible person in your previous organization who is not hostile toward you.

Duties. Some applications provide only a small amount of space in which to describe your past employment. Simply cover the highlights by using facts and figures to describe your responsibilities and achievements. Some examples might include:

◆ supervised a staff of five in an equine research facility

◆ assisted in 50 equine medical cases per month

◆ responsible for opening and closing the facility, and deposited $25,000 per week in the bank

Miscellaneous Information

Arrest Record. If you were charged but not convicted, our court system defines you as innocent, and therefore you do not have to indicate you are guilty of anything. You do not have to reveal any arrest record from a time when you were considered a juvenile.

Transportation. Some positions will require you to have your own car or to have a valid driver's license. If you are not in possession of a vehicle, merely write "I will obtain transportation if hired."

Licenses and Certificates. If you have any licenses or certificates that are job-related, be sure to mention them even if they are not officially requested on the application.

Volunteer Service. Some applications ask about volunteer experience. List those volunteer jobs that support your job objective. Consider placing any volunteer service that may be considered an important part of your overall experience under the work experience section of the application.

Hobbies. Indicate only those hobbies that support your job objective. For example:

◆ coached an average of 20 hours per week on various baseball teams for over 5 years

◆ traveled throughout Europe and am familiar with a diversity of cultures

Future Plans. Stress your interest in doing a better job through:

◆ specialized training

◆ specialized education

◆ hard work and performance

◆ career advancement

References

As stated in Chapter 5, the best references are responsible people who are familiar with your abilities and who think highly of you. You should consider such people as teachers, coaches, managers, and supervisors you know from previous positions; heads of organizations for which you do volunteer work; and any professionals with whom you have worked in prior positions.

Regardless of which persons you select as references, be sure to obtain their permission to list them as a reference and try to find out what they plan to say about you. It is often helpful to ask previous employers to write you a letter of recommendation in advance. You can then make copies of the letters to give to prospective employers when asked for references.

To assist you in practicing completing an application form, a sample form has been provided in this chapter (see Figure 6.2). The following is a list of tips on completing an employment application form.

Tips

1. Read and Follow Directions Carefully

Work slowly and cautiously. If the application asks you to print, do not write in cursive. If the application does not specify whether to print or write, it is best to print neatly. Most people print more clearly than they write. You do not want to misread a question and answer it incorrectly, forcing you to cross it out or write over it. It is acceptable to leave questions blank if they do not apply to you. To show the employer that you have read every question, you can answer questions that do not apply by writing "NA" for "not applicable" in the space provided.

2. Present a Neat Appearance

Neatness is important. A messy or smudged application form will give the employer a negative impression. This will get you screened and eliminated from the position quickly!

3. Be Prepared

Most employers will ask you to complete an application form at their place of business. If you always take your application worksheet with you, all you will have to do is copy the information from the worksheet onto the application form.

4. Bring Your Own Ballpoint Pen

Use a pen with black ink and a fine point. Usually the space provided on the application form is limited; therefore, a thick-tipped pen is not appropriate. A pencil allows you to erase mistakes but lacks professionalism. Always bring two pens in the event one becomes faulty. Some pens now come equipped with an eraser that makes it possible to correct mistakes neatly.

5. Be Honest

It is highly recommended that you do not attempt to falsify any information on the application. If you do so, you may be caught in a lie during the interview process, which will result in your being screened immediately out of the position. Even if you succeed in securing a position, a false statement on your application can lead to your dismissal at a later date.

6. Be Positive

Always maintain a positive attitude throughout your job search, and be positive when filling out an application. Application forms are designed to reveal negative traits about you. Anything negative will contribute to your being eliminated by the employer during the screening process.

APPLICATION FOR EMPLOYMENT

PERSONAL INFORMATION

DATE _____

NAME (LAST NAME FIRST)		SOCIAL SECURITY NO.		
		– –		
PRESENT ADDRESS	CITY	STATE		ZIP CODE
PERMANENT ADDRESS	CITY	STATE		ZIP CODE
PHONE NO. ( )	REFERRED BY			

EMPLOYMENT DESIRED

POSITION	DATE YOU CAN START	SALARY DESIRED
ARE YOU EMPLOYED? ☐ YES ☐ NO	IF SO, MAY WE INQUIRE OF YOUR PRESENT EMPLOYER? ☐ YES ☐ NO	
EVER APPLIED TO THIS COMPANY BEFORE? ☐ YES ☐ NO	WHERE?	WHEN?

NAME AND LOCATION OF SCHOOL	YEARS ATTENDED	DID YOU GRADUATE	SUBJECTS STUDIED
GRAMMAR SCHOOL			
HIGH SCHOOL			
COLLEGE			
TRADE, BUSINESS OR CORRESPONDENCE SCHOOL			

GENERAL

SUBJECTS OF SPECIAL STUDY/RESEARCH WORK OR SPECIAL TRAINING/SKILLS

U.S. MILITARY OR NAVAL SERVICE	RANK

FORMER EMPLOYERS
(LIST BELOW LAST FOUR EMPLOYERS, STARTING WITH LAST ONE FIRST)

DATE MONTH AND YEAR	NAME AND ADDRESS OF EMPLOYER	SALARY	POSITION	REASON FOR LEAVING
FROM / TO				
FROM / TO				
FROM / TO				
FROM / TO				

FIGURE 6.2
Application form

REFERENCES

GIVE BELOW THE NAMES OF THREE PERSONS NOT RELATED TO YOU, WHOM YOU HAVE KNOWN AT LEAST ONE YEAR.

NAME	ADDRESS	BUSINESS	YEARS KNOWN
1			
2			
3			

AUTHORIZATION

"I CERTIFY THAT THE FACTS CONTAINED IN THIS APPLICATION ARE TRUE AND COMPLETE TO THE BEST OF MY KNOWLEDGE AND UNDERSTAND THAT, IF EMPLOYED, FALSIFIED STATEMENTS ON THIS APPLICATION SHALL BE GROUNDS FOR DISMISSAL.

I AUTHORIZE INVESTIGATION OF ALL STATEMENTS CONTAINED HEREIN AND THE REFERENCES AND EMPLOYERS LISTED ABOVE TO GIVE YOU ANY AND ALL INFORMATION CONCERNING MY PREVIOUS EMPLOYMENT AND ANY PERTINENT INFORMATION THEY MAY HAVE, PERSONAL OR OTHERWISE, AND RELEASE THE COMPANY FROM ALL LIABILITY FOR ANY DAMAGE THAT MAY RESULT FROM UTILIZATION OF SUCH INFORMATION.

I ALSO UNDERSTAND AND AGREE THAT NO REPRESENTATIVE OF THE COMPANY HAS ANY AUTHORITY TO ENTER INTO ANY AGREEMENT FOR EMPLOYMENT FOR ANY SPECIFIED PERIOD OF TIME, OR TO MAKE ANY AGREEMENT CONTRARY TO THE FOREGO-ING, UNLESS IT IS IN WRITING AND SIGNED BY AN AUTHORIZED COMPANY REPRESENTATIVE."

DATE _____ SIGNATURE _____

INTERVIEWED BY _____ DATE _____

------------------------------- DO NOT WRITE BELOW THIS LINE -------------------------------

REMARKS

NEATNESS	CHARACTER
PERSONALITY	ABILITY

HIRED	FOR DEPT.	POSITION	WILL REPORT	SALARY WAGES

APPROVED: 1. _____ 2. _____ 3. _____
EMPLOYMENT MANAGER   DEPT. HEAD   GENERAL MANAGER

FIGURE 6.2
Application form (continued)

Telephone Techniques

*U*tilizing the telephone is one of the most efficient ways to look for a job in the horse industry. You do not spend a great deal of time traveling, and you get to speak to a large number of prospective employers in a relatively short time. The short-term goal of obtaining an interview can be achieved by three basic methods: a cover letter, a résumé, and a telephone call. As both the cover letter and résumé have been discussed in detail in previous chapters, let us now focus on the telephone as a vital tool in your quest for an interview and ultimately a career with horses.

Using the Telephone in Your Job Search Campaign

An applicant's first conversation with an employer will often take place before an interview is even scheduled. A conversation often takes place when the job seeker calls an employer to request an application or to inquire about job openings. You should realize that the telephone conversation, regardless of how short it is, plays an important role in the consideration of an applicant for a position.

Applying for a job by telephone is probably the fastest and most assertive method for landing a good job. This method is fast because you can quickly determine how many positions are available, when the application is due, and other important information. You may find it difficult to use the telephone at first; most people do. People feel uncomfortable in utilizing the telephone to contact potential employers for various reasons. Some people think it is too "pushy" to make a call and ask for an interview with a perfect stranger. Others fear they will be rejected, that they will seem nervous and be unable to carry on an intelligent conversation, or that they will be asked a specific question that they are unable to answer.

In order to overcome your fear of the telephone, it is important to practice. By practicing you will know what to say and be prepared for anything that may go wrong. Keep in mind that by utilizing the telephone in your job search, you are accomplishing the following:

◆ You are taking the initiative. This gives your contact the message that you are a self-starter.

◆ You become a real person. To the potential employer you are no longer only one résumé among many.

◆ You project a positive self-image.

The use of the telephone is a skill that everyone can learn. The basic steps in utilizing the telephone as a job-seeking tool are as follows:

◆ *Establish your objectives*—Be sure you are clear on your job objective in terms of the type of work you desire with horses, within a specialty, or in the chosen field you wish to pursue. Be sure to have a specific occupational title in mind when calling potential employers. Your goal is to obtain a personal interview with this specific occupational title in mind.

◆ *Establish a network of people to call*—Begin by establishing a network of people to call. This network might include family members, friends and neighbors, members of local horse clubs, past and present employers, members of a professional horsemen's association (e.g., United Thoroughbred Trainers of America), horse publications, barber or hair stylist, and local veterinarians. The Yellow Pages of your phone book will assist you in compiling your network. Refer to headings such as: Animal Hospitals, Horses, Horse Appraisers, Horse Boarding/Rental, Horse Breeders, Horse Dealers, Horse Trainers, Horse Transporters, Horse Shoers, Livestock Dealers, Feed and Grain Dealers, Riding Academies, Stables, Racetracks, Tack Shops, Saddlery and Harness, and Veterinarians and Veterinary Equipment.

You may obtain telephone directories in your immediate area through your local library. The Yellow Pages and business-to-business directories are available throughout the Untied States and can be easily purchased through any telephone company providing service to a particular area of the country.

After you have compiled your list for telephone networking and you have focused on a specific objective, you must decide how you will use your listed sources of telephone contacts. Begin by deciding where you wish to work. What part of the country appeals to you most? Consider the climate, the horse population, and the cost of living. Determine a specific area within the horse industry that interests you most. By doing this, you will become more focused in your job search. Visit your local library and ask to use the reference section. Ask the librarian for any business and agricultural guides. Do not hesitate to let the librarian know what you are doing and ask for any suggestions he or she may have to offer; for example:

> *I am researching career opportunities within the horse industry, and I would like to view some agricultural guides and veterinary directories that will provide me with the names, addresses, telephone numbers, fax numbers, and e-mail addresses of specific organizations.*

Telephone Script

Before you begin using the telephone to make contacts, develop something in writing to have on hand when you are on the telephone. Writing a telephone script will help you to present yourself in a positive manner. Write your telephone script in the same manner in which you naturally speak, and limit the script to the information a potential employer will want to know about you as well as a request for an interview. The script does not have to be a word-for-word statement, but it should be in a format that is easy to use. Your telephone script should include the following sections:

◆ *Introduce yourself*—State your name to the potential employer.

◆ *Say something friendly*—You might begin by saying something simple like: "How are you this morning?"

◆ *State your job objective*—It is best to begin with the following statement: "I am interested in obtaining a position as a (specific horse-related title)."

◆ *State your qualifications*—Briefly state your experience, education, and training as it relates to the specific horse-related position you are seeking.

◆ *State your personal traits*—You may want to say something like "I am dependable, a quick learner, and I am a hard worker."

◆ *Ask for an interview*—Sum up your telephone conversation by asking, "Would it be possible for you to spare a few minutes with me, at your convenience, to explore any opportunities in a personal interview?"

◆ *Confirm the date and time*—If you are granted an interview, simply state, "Excellent, that's Monday morning at 10:00 o'clock. Thank you, (Mr./Ms./Mrs. Name), I appreciate your taking the time to meet with me."

Once you have completed all the sections of your telephone script, you can develop a "telephone contact worksheet" (see Figure 7.1).

- **<u>NAME:</u>** (Hello, my name is _____.)

- **<u>FRIENDLY GREETING:</u>** (How are you this morning?)

- **<u>JOB OBJECTIVE:</u>** (I am interested in a position as a _____.)

- **<u>QUALIFICATIONS:</u>**

- **<u>PERSONAL TRAITS:</u>**

- **<u>REQUEST AN INTERVIEW:</u>**

- **<u>CONCLUSION:</u>** (Confirm Date and Time.)

FIGURE 7.1
Telephone contact worksheet

After completing your worksheet, keep revising your script until it sounds like you and feels comfortable. Be sure to write the script exactly as you will say it on the telephone. It is best to neatly type the script with double spacing to enable you to read it with ease while you are talking on the telephone. A typical telephone script might read as follows:

Good morning, (name of potential employer). My name is (your name), and my reason for calling is that I am interested in obtaining a position as a riding instructor. Let me tell you a few things about myself: I am a recent graduate of the Overland School of Riding, and I have taught English equitation, as well as basic dressage, for the past two years.

*I am dependable and quick to learn. Would it be possible for you to spare
a few minutes with me, at your convenience, to explore any opportunities
with your organization in a personal interview?*

Do not attempt to memorize your script or introductory statement. By doing
so, you provide additional pressure to your telephone call by trying to
remember what to say. Finally, ask a friend to play the role of a potential
employer and practice using your script. When you feel confident and familiar with your script, it will be time to put it to use.

Tips

1. Contacting the Person in Charge
*If you do not know the name of the person in charge, or you are not sure about
how to pronounce the name, ask the receptionist or the person answering the
telephone for the name or pronunciation. Be sure to get the correct spelling and
to use the name in your conversation.*

2. Contacting a Referral from Another Party
*In this instance, immediately provide the name of the person who suggested you
call. For example, say, "Hello, Mr. Green. Barbara White suggested I give you a call."*

3. The Best Time to Call
*There are good times and there are bad times to call a potential employer. The
following guide will aid you in deciding when to call:*

- *7 a.m. to 8:30 a.m.—**Bad time:** The employer may have just arrived and is
  having coffee or breakfast.*
- *8:30 a.m. to 9:30 a.m.—**Good time:** The employer should have arrived by
  now and be available to take calls.*
- *9:30 a.m. to 11:30 a.m.—**Bad time:** The work is in full force at this time
  and most people are busy. You will be interrupting their work.*
- *11:30 a.m. to 2:00 p.m.—**Bad time:** Everyone is at lunch during this time
  period.*
- *2:00 p.m. to 3:00 p.m.—**Good time:** The employer has returned from
  lunch and is usually available.*
- *3:00 p.m. to 4:00 p.m.—**Bad time:** The employer is busy wrapping up the
  day's activities.*
- *4:00 p.m. to 5:00 p.m.—**Bad time:** The employer's day is basically over and he
  or she may be tired.*
- *After 5:00 p.m.—**Bad time:** During this period of time, you would be
  infringing upon the employer's personal time.*

(continued)

Tips (continued)

It is important to realize that all horse operations are not on the same time schedule. For example, a breeding operation is very busy during the breeding season. Likewise, most training at racetracks is conducted between the hours of 5:00 A.M. and 11:00 A.M.; therefore, the previous schedule would not be applicable to a racetrack occupation. The following schedule would be more appropriate for making telephone calls as well as personal visits to a racetrack.

◆ 5:00 A.M. to 10:00 A.M.—***Bad time:*** Trainers and assistant trainers are busy getting their horses to the track for their daily exercise.

◆ 10:00 A.M. to 12 noon—***Good time:*** Horses have completed their daily exercise and are being fed their lunch. The potential employer will be relaxed and will be reflecting on the activities for the day.

◆ 12 noon to 4:00 P.M.—***Bad time:*** Most daytime racing is conducted during these hours, and the person in charge may not be available.

◆ 4:00 P.M. to 5:00 P.M.—***Good time:*** Trainers are usually back at the barn office at this time (if they are not racing a horse) to supervise the evening chores and feeding.

◆ After 5:00 P.M.—***Bad time:*** Everyone is tired at the end of a long day. The last thing they want to talk about is employment.

Note: *The day of the week is of little significance except that you should try to avoid calling on Fridays. Many employers alter their schedules on Friday to plan for the weekend activities.*

4. Be Enthusiastic

Your voice can make you sound friendly, intelligent, and dynamic. Keep your remarks positive. Make your calls while standing, as the standing position will force your voice to project an alert and enthusiastic attitude.

5. Be Flattering

Flattery conveys a certain amount of respect for the potential employer. For example, one flattering statement is "Thank you for taking time out from your busy schedule."

8 *Organizing Your Equine Job Search Campaign*

At this point you should have good insight into your interests, skills, and career values as they pertain to the horse industry. The next logical step is to organize and plan your search campaign. If you utilize a plan, it will make your job search in the horse industry easier and more successful.

Getting Organized

A good rule to follow is to organize your job search as if it were a job. First, you will need a place to work. Set up a job search office somewhere in your home. You will need the following items:

◆ *Basic equipment*—A table or desk, comfortable chair, telephone and answering machine, storage area and shelves, pens and pencils, note pad, 3 by 5 inch index cards, letter size envelopes, postage stamps, telephone books with Yellow Pages, calendar, and plan book.

◆ *Optional equipment*—Typewriter, word processor, personal computer, copy machine, and fax machine.

Keeping track of all your job search contacts can be a burdensome task. By utilizing 3 by 5 inch index cards, you can relieve some of the burden, as you can store important information about each contact who aids you in your job search. The example shown in Figure 8.1 illustrates the value of the job lead index card. Create one card for each prospective employer, and file the cards alphabetically according to the name of the organization, farm, or individual. The example shown in Figure 8.1 would be filed under *B*. Set up a second file for each day of the month, and number the cards 1 through 31. Once this is accomplished, you can utilize this file to file your index cards under the dates you desire to follow up on them. At the beginning of each week, simply review all the job lead index cards you find for that week. At the beginning of each day, pull the job lead index card filed under that date.

The Job Search Schedule

First, decide how many hours per week you plan to spend on your job search. Devote at least 15 to 20 hours per week to your job search if you are seeking a full-time position. The average job seeker spends about 5 hours per week looking for work. So it is easy to see that 15 to 20 hours per week is well above the average, thus increasing your chances of securing that special job within the horse industry you desire. Everyone works with a different timetable. Determine what works for you and stick with it.

Second, determine what days of the week you wish to devote to your equine job search. Most horse industry related businesses will be open Monday through Saturday. These are the best days for you to actively look for work.

Finally, decide what times you will utilize on each day you spend looking for work. For example, if you decided to spend 5 hours each Wednesday on your equine job search, you might delegate your time to begin at 9:00 A.M. and work until noon (3 hours), take a lunch break, and then work again from 1:00 P.M. to 3:00 P.M. (2 hours).

In today's competitive job market, especially within the horse industry, you must be knowledgeable in organizing actual skills and conducting a job search if you wish to secure the job you want. There are two ways to obtain any job. You can depend on luck and hope to stumble into the right job, or you can organize, prepare, and research in order to get the job you desire. Developing your résumé, preparing for interviews, networking, targeting your cover letter, and following through are the skills you need to succeed.

There is a lot more to securing a position with horses than just writing a good résumé, visiting farms or veterinary hospitals, and filling out application

---

NAME OF ORGANIZATION: Beaumont Arabian Farm

ADDRESS: P.O. Box 135, Beaumont, Wisconsin 06517

CONTACT: Jane Doe, Manager   **PHONE NO.:** (525) 123-4567

SOURCE OF LEAD: Aunt Sarah

NOTES: 6/12 called Ms. Doe, on vacation. Call back on 6/18. Called on 6/18—Interview appointment on Tues. 6/20 at 1:00 p.m. On 6/20 Ms. Doe toured facility with me and conducted interview. Sent thank-you note same day. Call back on 6/30. Second interview with farm owner on 7/6 at 9:00 a.m.

FIGURE 8.1
Job lead index card

forms. Most people seeking a job within the horse industry will spend months looking for a job because they are totally unaware of the fact that there is a strategy that can be used to locate jobs. A self-initiated job search increases your overall awareness of the entire horse industry, places you in contact with professionals in your field, and decreases the amount of time it takes you to secure a job and find that special position within the horse industry that you desire.

By utilizing your job search tools, you will obtain in-depth knowledge of the horse industry, increase your professional contacts, and increase your resourcefulness. All of this will eventually lead to job satisfaction because you will be seeking a job *you* want.

Tools for the Job Search

There are many effective tools available to find a position within the horse industry. Some of the most common tools are: classified advertisements, mailing your cover letter and résumé, help-wanted signs, and employment agencies specializing in the horse industry. These tools are fine, but you usually have to wait for things to happen. Waiting for a response from a potential employer will not aid you in finding the job you want.

The most effective job search tools allow you to take action in determining your own destiny in the horse industry job market. These tools may be generally referred to as the "uncommon" tools. Some examples of uncommon job search tools include: networking, informational interviewing, and utilizing the Internet. If your job search plan combines both common and uncommon job search tools, you are in a better position to achieve your goal of finding that job within the horse industry for which you are best suited.

COMMON JOB SEARCH TOOLS

The Library

A library is an excellent source of job search information. The library is also a good source of major horse trade publications which carry a large number of help-wanted advertisements. For a complete list of horse publications, refer to the On-line Resource™ that accompanies this text (explained in more detail in the Preface). The reference section of the library will provide you with career guides as well as information on how to find and use job search materials. Telephone directories, local and out-of-town newspapers, and personal computers linked to the Internet may also be available, free of charge, at the library. Finally, the library provides the perfect setting for conducting your research.

Commercial Bookstores

Most bookstores offer a wide selection of books on all phases of the job search. Some bookstores may even have a job search section available for their customers.

Classified Advertisements

Less than 20 percent of all jobs in the United States are filled as a result of a classified advertisement. Answering classified advertisements, however, is an important part of your equine job search campaign. Do not limit the number of advertisements you answer. Requirements for a horse-related position are usually exaggerated in a classified advertisement. If you feel you can handle the position offered, answer the advertisement and state your reasons in your cover letter.

Begin by researching your local newspapers or, if you desire employment outside your area, obtain copies of the Sunday edition of a newspaper from the area in which you wish to become employed. The best place to look for help-wanted advertisements is in the horse industry trade publications, but another source of help-wanted advertisements can be found in the Sunday classified section of a major newspaper. You will find that Monday and Friday editions of newspapers contain the smallest selection of help-wanted advertisements.

It is recommended that you do not answer an advertisement on the same day as it appears. If you send your cover letter and résumé on the same day as the advertisement appears, the prospective employer will receive them at the same time he or she is overwhelmed with mail. Wait a few days and then send your material, as this allows the prospective employer to pay more attention to your résumé and cover letter. The best day to answer an advertisement that appears in a Sunday edition of a newspaper is the following Tuesday. Your cover letter and résumé will stand out from all the rest. If you happen to see an advertisement after it has first appeared, respond anyway; you have nothing to lose.

RULES FOR ANSWERING CLASSIFIED ADVERTISEMENTS

1. The advertisement should tell you what the potential employer does (sell a product, perform a service, and so on).
2. Is the title of the job listed in the advertisement? The title is needed to determine your suitability.
3. Is the salary range listed in the advertisement? This information is needed to determine if the salary range meets your requirements.
4. Avoid any short advertisements (four to five lines). Why seek employment with an organization that fails to purchase adequate space to describe a job opening?

By reading the classified help-wanted section on a daily basis, you will begin to see trends in the horse industry job market. For example, if you noticed a large number of "horse farm manager" advertisements over a period of time, you could safely say that farm managers are in demand. Knowing that, you could be more selective in choosing a farm manager position.

Finally, in order to increase your chances for a positive response from a classified advertisement, follow-up is necessary. One week after sending your original cover letter and résumé, send a follow-up letter containing additional information. Be sure to mention the fact that you have already sent your résumé.

Employment Agencies

If you decide to utilize the services of an employment agency, bear in mind that there are different types: public, private, and temporary. Regardless of the type of agency you consult, however, be sure that you fully understand the service provided, the extent of your involvement, and your financial commitment.

Public Employment Services. There are many nonprofit public employment service agencies. Educational institutions such as colleges, universities, community colleges, and high schools employ counselors with professional qualifications. Government agencies such as the Veterans Administration and the various state employment services provide career guidance services. The Department of Labor, in conjunction with the state departments of labor, provides a large system of public employment agencies known as Job Services, Employment Security Offices, or Manpower Services. These agencies receive funding from the states and have access to employment data throughout the United States. Potential employers list horse-related positions with these agencies, and job placement counselors attempt to locate suitable applicants for the positions listed. To utilize these services, call your local office and make an appointment to register. Registration requires that you provide your work history, career goals, and so on.

Private Employment Services. There are many reputable private employment services that provide excellent services for the job seeker pursuing employment within the horse industry. However, there are many private employment agencies that are guilty of making unsubstantiated claims about their success and that require a large registration fee from the job seeker. All private employment agencies must be licensed by the state and counties in which they operate. The job seeker may contact the County Clerk's office or the state Attorney General's office to determine the credibility of a particular employment agency. These officials will provide you with any records pertaining to complaints or legal action taken against any licensed employment agency. Unlike the public employment services, the private agencies provide their services for a fee. In some cases, the potential employer pays the required fee. If you are required to pay for this service fee, as a job seeker you will no doubt be asked to sign a contract at the time you register with a particular employment agency. It is important that you read the contract carefully before you sign it because it will state your financial obligation. Do not hesitate to research the reputation of any agency with the proper regulating authorities before you decide to utilize the services of a private employment agency.

Temporary Employment Agencies. The temporary job market within the horse industry is one avenue of employment that should not be overlooked by the active job seeker. Temporary jobs with horses do turn into permanent jobs. Many employers within the horse industry are reluctant to hire full-time employees because they are concerned about fringe-benefits costs. Temporary employment agencies provide a variety of temporary workers for the horse industry; the jobs range from grooms to professional accountants. Some temporary agencies are willing to provide temporary workers with specialized training, which can prove to be most beneficial to the person who is moving from one field to another.

The temporary job seeker designates the total number of hours he or she wishes to work, and the temporary agency does the rest. However, the agency does not provide any fringe benefits to the part-time worker. Temporary workers are usually hired to fill an unexpected vacancy. Those who do well may be the first candidates to be considered to fill the vacant position.

Professional Horse Associations

Professional horse or trade associations are an excellent source for your job search. Most professional horse associations publish a monthly newsletter or magazine that contains available job listings. As a member of this type of horse organization, you are able to list your membership on your résumé. This relays to the potential employer that you are active in your field of interest. A professional horse association can also provide you with membership lists or directories you can utilize to develop new contacts in your job search.

Educational Placement Offices

Most colleges and universities offer job placement services to their alumni. Vocational schools specializing in a particular area of horsemanship, both private and public, offer job placement services for their graduates as well. If you are a college graduate with a degree in Equine Science or a related field, you would be wise to take advantage of the services offered by the institution's placement service.

Breed Conventions and Equine Trade Shows

An equine trade show, horse industry conference, or annual breed convention is a perfect setting for making direct contact with prospective employers. The equine trade show, convention, or conference represents an unusual job-hunting opportunity and should not be overlooked. Attend these events with your business cards and copies of your résumé. Make it a point to visit every booth that looks promising. Leave a business card or your résumé if the opportunity arises. Get as many names, addresses, Web sites, and telephone/fax numbers as possible. Collect any and all information that is available at each organization's booth. After the event:

◆ Decide on the best possible way to follow up. Ask yourself: Should I call or write?

◆ Organize your notes in the order in which you will contact each prospective employer. What organizations seem to offer the most promise? Who are the people you must contact?

◆ Adjust your résumé to fit specific jobs. Emphasize those skills you feel each of the prospective employers is seeking.

◆ Begin writing letters expressing your interest or make telephone calls if that is more suitable. If you decide to write, be sure enclose a résumé.

◆ If you plan to make contact with organizations that do not fit your credentials, write letters and enclose a résumé. State that you would be interested in working with them if jobs develop in the future.

◆ Establish follow-up dates.

Direct Mail Campaign

This method of seeking employment is the most common method of marketing one's experience, skills, and accomplishments. With this method, you utilize an extensive mailing list of potential employers in your field of interest and mail cover letters as well as your résumé to all of them. The rationale behind this method is the more résumés that go out to prospective employers, the better the odds of finding a job. Before considering this method, however, you should take into account that the average success rate of a direct mail campaign is approximately 2 percent.

The reason the success rate of a direct mail campaign is not as high as that of a targeted mail campaign is that it is not focused. With the direct mail method, you are forced to send a general form letter. Another disadvantage is that mass mailing can become quite costly to the job seeker. If you wish to do a résumé mailing, use a targeted direct mail campaign.

The four key elements of a successful direct mail campaign are:

1. the correct prospective employer list
2. your cover letter and résumé
3. the personal interview
4. the follow-up

Setting up a Mailing List. In order to set up a mailing list, you have two basic options: put a mailing list together by yourself, or purchase a preexisting list from a horse industry publication. In order to compile your own list, use your local library as a starting point. Your main source will be reference directories and trade publications. Use a separate index card for each prospect, and record the following information about each target horse-related organization.

◆ name, address, and telephone/fax number of organization

◆ person in charge (manager, owner)

◆ services or products

◆ location and general information

If you can afford the expense, purchase a preexisting mailing list, as it will save you valuable time and effort. Most trade publications sell their subscriber lists. These lists are ideal for a direct mail campaign, as they are aimed at your targeted group of prospective employers.

Preparing Your Cover Letter and Résumé. Your résumé will naturally accompany your cover letter in a direct mail campaign. The letter you construct for a direct mail campaign is called a "broadcast letter," as it broadcasts your availability to a large audience. The broadcast letter must be designed in a way to appeal to the greatest number of prospective employers as possible (see Chapter 4). The ideal aspect of this method is that you only have to write the letter once. That one letter can then be duplicated and mailed to the entire mailing list. Employers prefer letters that address their specific needs and desires, not some general "To whom it may concern" letter, so you must construct your broadcast letter to read as though it were written for an audience of one and not hundreds. Try to personalize your letter by focusing it as much as possible. Focus on your job objective by discussing your area of interest and what is happening in the industry, and then state your qualifications. For example:

> *As the number of racetracks are declining each year, the role of the marketing director is becoming more and more challenging. Public relations has become a vital function and an area in which I have special expertise.*

The remainder of your letter should be the same as any other cover letter that you have written. (See Chapter 4 for specific letter formats.)

Make follow-up telephone calls two weeks after your initial mailing. You can increase your success rate to 10 percent, which is far better than the normal low 2 percent rate. In your follow-up call, refer to the date you sent the letter, state your job objective, and request an interview. If you are not landing any interviews, consider changing your letter or expanding your mailing list.

Following are some *do's* and *don'ts* to consider in your direct mail campaign.

DON'T

- ◆ Do not send more than one copy of your letter to the same organization.
- ◆ Do not send photocopies or an obvious form letter. Although this is a large mailing, you should downplay this fact as much as possible.

DO

- ◆ Use a computer to print your letters, as each letter will appear to be individually prepared.
- ◆ Match the quality and color of the paper and envelopes you use for the résumé and letter.

UNCOMMON JOB SEARCH TOOLS

Networking

This process enlists everyone you know to aid you in your job search. Networking focuses on building a pool of personal contacts who can provide you with career information that can ultimately lead you one step closer to getting hired. Networking is a shortcut to finding the information you need by questioning your personal contacts for advice and input. Spend time developing a circle of contacts who are aware of the type of position you desire and what type of job you can perform.

Begin developing a network list by including friends, relatives, teachers, classmates, neighbors, members of social clubs, sport clubs, church groups, and associates. Include people with whom you are in close contact on a regular basis, such as relatives of friends or friends of friends, even if you do not know them well. Your goal is to schedule a meeting with each person on your list in order to obtain as much information as possible and to advise them of your skills and availability for employment.

Start with contacting the people you know well first, and then move on to the others. Remember that the more people who are aware of your needs and skills, the better are your chances of finding someone who knows of a job opportunity. Always stay in touch in a polite and friendly manner with everyone on your list.

Computerized Research

Utilizing an on-line personal computer for a job search in the horse industry is actually no different from any other method of job seeking. The advantages of using a computer is that it allows you access to updated specific openings throughout the United States and allows potential employers to scan your résumé quickly. To best utilize a personal computer in your job search, focus on two basic types of job banks—résumé banks and job openings:

◆ *Résumé Banks*—Electronic résumé files maintained by a firm that allows prospective employers to review the files. The cost for this service is minimal, and some educational institutions offer résumé banks for free.

◆ *Job Openings*—Electronic classified advertisements that you can view with a computer. These electronic advertisements are very specialized in nature and focus on a specific area within the horse industry.

Note: Both of these types of job banks are available through most on-line services.

The Internet is an excellent tool for the job seeker. It provides job and career information for the horse industry. On-line services also provide a variety of services offering employment-related information for the job seeker. The Internet has something for everyone, from the recent graduate to the seasoned professional.

Listings found on-line cover virtually every industry including the horse industry. The jobs that are posted on-line are usually listed by industry, location, and occupation or a combination of these. Utilizing the Internet in your job search is a lot easier and provides more focus than dealing with newspapers and trade publications. The job seeker, however, should be aware of the costs of using on-line services; charges vary from service to service. Most on-line services charge a monthly fee and then an hourly fee for actual use. Check out the on-line services at your local educational facilities, library, or state employment offices first, as these may be available to you free of charge.

Most on-line services offer job and career-related information. It is best to select the service that meets your budget and job search needs. The following are some of the main on-line services that are available:

Compu Serve Inc.
5000 Arlington Centre Boulevard
Box 20212
Columbus, Ohio 43220
(800) 848-8199

Prodigy Services Company
445 Hamilton Avenue
White Plains, New York 10601
(800) 776-3449

America On-Line, Inc.
8619 Westwood Center Drive
Vienna, Virginia 22182-2285
(800) 827-6364

There are many interesting horse industry Web sites available to the on-line subscriber. Once you are on the Internet, start at an equine link site. This initial site will provide you with a catalog of equine Web sites by topic. In order to reach any of these Web sites, you merely click on its name.

COMMON EQUINE WEB SITES

- ◆ American Horse Council—http://www.horsecouncil.org
- ◆ American Veterinary Medical Association—http://www.avma.org
- ◆ Cybersteed—http://www.cybersteed.com
- ◆ Equine School and College Directory—http://www.hhyf.org
- ◆ Equinet—http://www.equinet.com
- ◆ HayNet—http://www.freerein.com/haynet/index.html
- ◆ Horse Interactive—http://www.thehorse.com
- ◆ HorseWeb—http://www.horseweb.com
- ◆ NetVet—http://www.netvet.wustl.edu/horses.htm
- ◆ Oklahoma State University's Horse Resources—http://www.ansi.okstate.edu/library/equine.html

EQUINE JOB SERVICES

- ◆ Equimax—http://equimax.com
- ◆ Equistaff—Jobs@equistaff.com
- ◆ Horse Quest—www.horsequest.com

- Professional Equine Employment Management, Inc.—
  http://www.equinepros.com
- Department of Agriculture, Cooperative Extension Job Vacancies—
  gopher://sulaco.oes.orst.edu/11ext/jobs
- Department of Agriculture, Job Bank Bulletins—almanac@esuda.gov

Informational Interviews

An informational interview is an invaluable tool for increasing the number
of your professional contacts and your insight into the horse industry.
During this phase of your research, focus on getting the feel of the horse
industry or the job itself. The information you gather can save you a great
deal of time by giving you a true picture of your dream job. The following are
all part of the informational interview process:

1. Contact people on your network list. Most people will know someone
    you can speak to, or perhaps a friend who does. Be sure to ask for the
    contact's name, telephone number, and official position. Call and make
    an appointment to meet to discuss jobs. It is important that you stress
    the fact that all you seek is information.

2. During the informational interview, ask your contact about the pros and
    cons of his or her job or career. Some of the basic questions you should
    ask are:

    - What type of job opportunities are available?
    - What type of person fits these job opportunities?
    - Are there advancement opportunities available?
    - Would you choose this profession if you could do it over again?
    - What advice can you give me about getting into the horse business?
    - What type of training or education and skills do I need?
    - Can you provide names of other individuals to talk to, either in this
      organization or the horse industry in general?
    - Do you feel I have the right background, qualifications, and experi-
      ence for any specific occupation in the horse industry?

3. Be sure to obtain other referrals. You will need information from at least
    three people in the field in order to get a balanced view and to build up
    your network of contacts.

4. Your goal is to get the contact interested in you without actually asking for
    a position. Whatever you do, do *not* ask for a position during an informa-
    tional interview. Instead, mention your job search and your need for
    names, information, and specific advice. Get the referral interested in you.

5. Summarize your objective. Although you have already outlined the infor-
    mation you are seeking in your initial telephone call to your referral,
    repeat it at the beginning of your interview.

6. Do not underestimate the importance of your informational interview. Although it is an informational interview, it is still an interview. Treat it as one. At some point in time this referral will be in a position to hire or recommend someone for a job. Naturally you want that person to be you, so you should do the following:

    ◆ Act as though you were being interviewed for a position.

    ◆ Be confident and positive about your background and goals.

    ◆ Most of all, express a genuine interest in your job search and in the interview itself.

7. After each informational interview with a valuable contact, always take the time to send a short thank-you note. Informational interviews are similar to standard interviews. You would send a thank-you note to a prospective employer, so you should do the same for referrals and contacts. It is an excellent way to remind them of your meeting and you.

Maintaining Your Job

*I*n this chapter we will discuss the various techniques that can help you keep your job once you get it. In addition to maintaining your job, we will also discuss the methods involved in terminating your position with your present employer, for whatever the reason.

Perhaps things are not going as smoothly as you anticipated after you begin your new job. Maybe you are confused about what is expected of you. Everyone around you seems like a pro and you begin to feel inexperienced. It is quite normal to be apprehensive, especially if you are in your first job.

The First Job

Let us focus on the first job situation for now. Your first job is important in that it lays the groundwork on which you will build your career with horses. Your first job will enable you to develop your first horse industry contacts. Not only will you be observing those around you, but your employer will be observing you to determine your suitability for the job.

You will begin to meet people inside and outside your organization who may be able to help you advance your position or help you get your second or third job. You will develop a network of people in whom you can confide and ask for information and advice. Novice employees in the horse industry are usually unsure of their abilities. Therefore, the first job can be a boost to your self-confidence. Another bonus that the first job offers is your first opportunity to obtain an inside view of your chosen career with horses.

It is one thing to read about a profession in the horse industry and another to actually be a part of it. The first job can help you measure your aptitude and interest in a particular career with horses. Perhaps more importantly, the first job will aid you in developing a broader view of the way professionals and organizations function. By observing the mistakes and successes of others, you will develop a style of your own.

Obviously, your first job is important in many ways; however, it is highly unlikely that it will determine the course of the rest of your life. Most people change jobs and even careers several times in a lifetime, and probably so will you.

Now that you have accepted an offer and are ready for that first day on the job, you have made a commitment to give your best efforts to your new employer. This means that you will be required to learn everything about the job as quickly as possible. The following are tips for those first few days.

Tips

1. Be On Time for Work Every Day
Punctuality is a major management consideration when it comes to promotions and salary raises. It is also one of the major causes for discharge, especially for new employees.

2. Follow Directions Without Question
Wait until you learn more about the organization and your responsibilities before you challenge orders. Perform your duties correctly, and a time will come when your opinions will be welcomed.

3. Try to Get Along with Everyone in the Organization with Whom You Come into Contact

4. Avoid Joining Cliques

(continued)

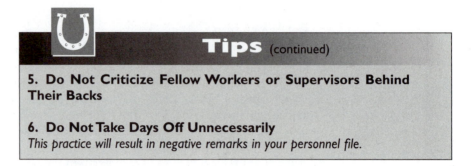

Tips (continued)

5. Do Not Criticize Fellow Workers or Supervisors Behind Their Backs

6. Do Not Take Days Off Unnecessarily
This practice will result in negative remarks in your personnel file.

Remember, regardless of how good or bad your first job is, you can walk away from it with valuable knowledge about yourself, your work, and the realities of life that your education and previous experiences never taught you.

Becoming a Valuable Employee

There is more involved in performing your role as an employee than collecting a paycheck. Achieving company goals, being committed, and producing goods and services are the three basic characteristics of a valuable employee. Understanding the needs of the business can make you a more effective and valuable employee. Basically, employees who see better ways to achieve organizational goals become more valuable to the overall operation of the organization. As an organization grows, so do the employees' opportunities for change and growth. Businesses depend upon employees to be committed to the goals of the business.

The term *indispensable* refers to a sense of trustworthiness, reliability, and skill that persuades other people to come to you when there is a tough problem to solve. An indispensable person is one about whom it is often said, "I wish I had a dozen like her (him). We'd get the job done in half the time!" You can do a great deal to make yourself so important to the day-by-day operation that your employer would be the loser if you were to leave for any reason. The following actions will aid you in reaching that indispensable status. These same actions may also increase your prospects in the organization—put you in line for salary increases, promotions, and so on.

Strive to Be the Best.   The first and simplest action is to master your own position. Learn as much about your position as possible. Do not hesitate to put in as much overtime as needed, not only for the extra income, but to increase your total of operating hours and experience.

Master a Related Specialty. Search for tasks or areas of responsibility that, while not directly a part of your job, allow you to become active in peripheral areas. This will enable you to gain the attention and high reward that can be crucial to your survival within the organization. For example, if you are employed as a veterinary assistant, your position may stress duties that pertain to surgical procedures. However, it would be to your advantage to learn some basic computer skills and master the billing procedures performed by the office staff. Naturally you should go about such tasks without offending any of your fellow staff members.

Become a Specialist in the Organization. In many organizations there are opportunities for alert and interested employees to develop special positions for themselves. Some possible areas in which you can become a specialist are as follows:

- Learn the history of your employer and the physical area in which your place of employment is located.
- Become knowledgeable about the community in which your employer is located.
- Become knowledgeable about the horses and their offspring that are associated with your employer.

Be Aware of New Developments. Often an organization decides to expand by, for example, adopting a new sales venture or marketing strategy. It may be an entirely new area that takes the organization into unfamiliar aspects of the horse industry. Sometimes such expansion will result in a second location for operations or perhaps a new department. Each innovation, each expansion of this kind, creates opportunities. People are needed to supervise and to operate the facilities. Employees who are alert for such developments can improve upon their chances for advancement considerably.

Learn Something Crucial. Every horse-related operation has aspects of its business that are sensitive, that are not for public knowledge. One way of making yourself a difficult employee to fire is to work yourself into a position that gives you access to this vital material. For example, if you are employed by a public riding stable, it would be to your advantage to become knowledgeable about the legal liabilities of the business. You could, for example, become familiar with workers' compensation laws in your state and learn about the necessary insurance forms, legal contracts, and building and zoning codes in your area of operation. By knowing such special information regarding your employer's operation, you automatically become a key person.

Make a Major Contribution to Profitability. No matter where you are in the organization, you undoubtedly can improve operations, cut costs, and make other worthwhile contributions by making constructive suggestions.

The kind of ideas that can make a major difference in your situation are basic ones that will score an obvious improvement in profitability. There is no limit to the aspects of organizational activity that can benefit from a sound idea. Some of the possibilities include new products and services, improvements in basic methods of operations, and new markets.

Qualities of Good Employees

The typical employer uses a range of qualities to evaluate employees. In general those who rate high on these desirable qualities have a good grip on their jobs. Figure 9.1 lists examples of personal characteristics found in employees who are successful.

There is a wide range of things you can do in the course of your day-to-day work that can make you shine in the eyes of your employer. With these qualities working for you, your chances of maintaining your job are excellent.

Job Skill. Regardless of your position in the horse industry, your daily work represents a stage at which you can show your stuff. In some cases, an employee is one of a group of people doing approximately the same kind of work. Let us say, for example, that you are a riding instructor at a riding school and that there are three other instructors at your level. To some extent a measure of your job skill is represented by questions like these:

◆ How good are you compared to your competitors?

◆ Are you the best of the four riding instructors?

◆ Are you somewhere in the middle?

There are two aspects of making an employer value your job skills. One is to master them as best you can. The other is to make the moves that will tactfully build your reputation to match your performance—for example, letting your views be heard at meetings or discussions about the work, talking to your boss and colleagues about some of the finer points of the tasks that arise, and voicing opinions and backing up your point of view with authority.

Acceptance of Employer's Goals. The employee who is aware of the employer's objectives and signals the acceptance of these objectives can strengthen his or her position in the eyes of the employer. Today it is an accepted fact that employers and employees have many objectives that are identical. The survival of the organization, its growth, and for that matter the increased capability of the employees are mutually desirable goals.

Accuracy	paying strict attention to details to see that work is correct
Ambition	having a desire to achieve or accomplish a task or goal
Aptitude	having a natural talent or ability
Articulation	being able to express ideas clearly
Attitude	your feelings toward someone or something
Confidence	being comfortable and self-assured with yourself
Consideration	being thoughtful and kind to the feelings of others
Cooperation	getting along and working efficiently with others
Courtesy	being kind to fellow workers, associates, and the public
Dependability	following through on responsibilities and commitments
Diligence	working hard toward achieving a goal and not giving up
Efficiency	doing something quickly and thoroughly
Empathy	putting yourself in another's position to feel and understand his or her feelings and position
Enthusiasm	having a zest for work and life
Ethics	being honest and knowing right from wrong
Flexibility	being able to change when a situation requires change
Friendliness	being warm and interested in others
Honesty	acting in an ethical and trusting manner
Imagination	being able to think of new and creative ideas
Independence	making your own decision
Initiative	being able to identify work to be done and complete it without direction
Leadership	taking responsibility for guiding and planning the activities of a group
Loyalty	being faithful and not demeaning your employer or place of employment
Optimism	looking on the positive side of things
Productivity	the amount of work you produce
Punctuality	being on time
Respect	treating others with consideration
Sincerity	being truthful in your relationships with others

FIGURE 9.1
Personal characteristics of successful employees

Cost Consciousness. The employee who treats the employer's property as carefully as if it were his or her own, who respectfully spends the employer's money for equipment and materials, takes on a special value to the employer.

Ideas and Problem Solving. No matter what kind of job you have, no matter what type of work you do in the horse industry, there is something else you can produce in connection with your work that will make you look good to your employer. It is simply this: Come up with ideas; come up with solutions to problems confronting the day-to-day activity of the organization.

Work Improvement. One of the ongoing efforts of every organization involved in the horse industry is the improvement of work methods. Usually what happens is an employee figures out a way to do the work in a somewhat more efficient manner. A farm manager may develop a method of handling the incoming mail that allows him or her to spend more time on other farm duties. Your interest in improving the work that you are doing or that is part of your department's operation is another positive quality that your employer will view with favor.

Cooperativeness. No one can force you to be cooperative on the job unless you really feel that way. Many people mistakenly feel that just by doing their job, they are earning their salary and therefore being as cooperative as possible. There is a level of participation, however, that functions beyond the "get by" level. The employee for whom going a bit above and beyond the call of duty is no problem is the one who usually shines in the eyes of the employer.

Growth Potential. Every employee has the potential for growth on the job. In some cases this means improvement in present job skills. In others it means acquiring new abilities. Growth on the job does not necessarily mean moving up the ladder; it can also mean becoming a more valuable employee by virtue of increased experience. The most desirable type of experience is the job know-how that people pick up on the job. Employers cannot help being impressed by employees who make an effort to acquire new information and to conduct themselves in a manner that is meaningful to their employers.

Job Termination

It is important to discuss the subject of job termination, as there are no guarantees that you will not be fired from your position. It can happen to anyone. No one is completely indispensable. However, there are a few warning signs that may aid you in holding on to your job.

WARNING SIGNS THAT YOUR JOB MAY BE IN JEOPARDY

1. Your employer has started treating you differently. Your efforts do not seem to be recognized or appreciated.
2. Your subordinates are also treating you differently. They go around you or above you if they have a complaint or need direction or information.
3. Your job is getting too big for you to handle alone.
4. Your job is getting smaller and before long you will not be needed at all.

Job security has a completely different meaning today than it did 20 years ago. In the past, individuals worked for single employers and followed a fixed career path, with periodic promotions and salary increases. Employees felt that they had a loyalty to their employers whom they considered as their benefactors. Today, employees can no longer wait for opportunities to come to them; they must take the initiative, or they may find themselves unemployed. It is very difficult to find a new job while you are still employed. However, it is successfully done all the time. The following are some tips on how to balance your efforts in finding a new job with your responsibilities in your present job so that your job search activities will not be detected.

Tips

1. Dealing with Your Present Employer
It is difficult to act naturally around a current employer while talking to prospective employers. Try to maintain your regular lunch schedule. Long lunches occurring too frequently are a dead giveaway. Keep you personal calls to a minimum on your present job. You may inadvertently discuss future employment. Never divulge your intentions to fellow workers, no matter how friendly you are with them. Be sure that your relations with your superiors remain normal.

2. Scheduling Interviews
Avoid interviews with prospective employers during working hours. It is best to schedule several interviews for a single day and take a sick or vacation day.

3. Leaving Your Present Position
Once you find a great new job, your next challenge will be to break the news to your present employer. There is basically only one responsible way to leave your present job: give proper notice of your intentions, express your appreciation for the experience, make arrangements for an orderly transition, and say good bye in a pleasant manner.

STEPS FOR LEAVING YOUR JOB GRACEFULLY

◆ Determine how much notice to give. Two weeks is standard. For a higher-level position, a longer period of time would be appropriate.

◆ Schedule a personal meeting with your employer or your immediate superior. Prepare a positive resignation speech. It should simply express your intention to leave, date of departure, and appreciation for the opportunity to work for the organization.

◆ Offer to train your replacement. If no replacement has been selected by your last day, volunteer to be available by telephone for a week or two after you have left.

◆ After your resignation meeting, prepare your written resignation. Address it to your employer. Do not elaborate on your reasons for leaving. (See Figure 4.7 in Chapter 4)

◆ Be aware of the possibility of counteroffers. If you are tempted by one, review the reasons you decided to leave. If they are still valid, proceed with your plan of action and politely decline any counteroffer.

◆ Handle yourself professionally and responsibly at all times. Resignations can cause hard feelings. Never burn your bridges behind you. You may someday need references, networking contacts, or information.

Remember, if you act like a mature adult and a professional when leaving an employment position, the chances are good that your employer will treat you in the same manner.

10

Equine Occupations

T he purpose of this chapter is to provide the reader with detailed information about the various occupations available within the horse industry. The occupational profiles are written in broad general terms for those exploring horse career opportunities as well as those presently involved in a job search campaign. The equine occupational titles have been arranged alphabetically as a matter of convenience for the reader (see Figure 10.1).

Each equine occupational title contain vital information in the following categories:

- ◆ Description
- ◆ Educational and Training Requirements
- ◆ Personal Qualifications
- ◆ Licenses and Certification
- ◆ Employment Outlook
- ◆ Related Occupations

Salary or wage information for specific equine occupations is not presented as part of the occupational descriptions. Recent studies have illustrated that salaries for various levels of employment vary significantly among the different states in the union. In order to obtain salary information on a specific horse industry occupation, consult the local Department of Labor Office, and talk to individuals who are presently employed in the occupation of your choice. Those occupations at the professional and managerial level usually command higher salaries than technical and entry-level positions. Finally, salaries are commensurate with qualifications and experience.

Equine industry experts estimate that the horse population will increase in excess of 10,000,000 animals in the near future; and increases are expected to be higher each year. Therefore, the occupational outlook for many occupations in the horse industry is positive.

Animal Control Officer	Farm Equipment Dealer
Announcer	Farm or Ranch Manager
Appraiser	Farrier
Architect	Feed and Grain Dealer
Assistant Horse Trainer	Foaling Attendant
Auctioneer	Groom
Auction Sales Employee	Groundskeeper
Bloodstock Agent	Harness Horse Driver
Breeding Technician	Hippotherapist
Clocker	Horse Breeder
Equine Accountant or Bookkeeper	Horse Industry Secretary
Equine Advertising Specialist	Horse Insurance Agent or Broker
Equine Artist	Horse Insurance Claims Adjuster
Equine Attorney	Horse Insurance Underwriter
Equine Book Dealer	Horse Show and Event Judge
Equine Chiropractor	Horse Trailer and Van Dealer
Equine Consultant	Horse Trainer
Equine Dentist	Horse Van or Truck Driver
Equine Educator	Hot Walker
Equine Extension Service Agent	Identifier
Equine Fence Dealer or Installer	Jockey
Equine Geneticist	Jockey Agent
Equine Journalist	Jockey Valet
Equine Librarian	Laboratory Animal Technician
Equine Loan Officer	Mounted Park Ranger
Equine Massage Therapist	Mounted Police Officer
Equine Nutritionist	Outrider
Equine Photographer	Pedigree Researcher
Equine Real Estate Broker and Agent	Pony Boy or Girl
Equine Recreation Director	Retail Tack Shop Operator
Equine Sportscaster	Riding Instructor
Equine Transportation Specialist	Saddlesmith
Equine Travel Agent	Starter and Assistant Starter
Equine Veterinary Assistant	Veterinarian
Equine Videographer	Veterinary Pharmaceutical Salesperson
Exercise Rider	

FIGURE 10.1
Equine Occupational Titles

ANIMAL CONTROL OFFICER

Description

Animal Control Officers investigate abuse complaints concerning horses and other animals. They may be involved in the rescue, capture, or administration of first aid to horses and other animals. They have the legal authority to make arrests and seize property in cases involving animals, and they have the power to issue a citation to offenders. The Animal Control Officer may also be required to lecture on humane treatment of animals and educate the public on animal needs and behavior. Finally, Animal Control Officers may be involved in inspection and follow-up calls at zoos, riding stables, circuses, and animal theme parks.

Educational and Training Requirements

A high school education is mandatory for this position. A two- or four-year college degree in Equine Science, Animal Science, or Law Enforcement is beneficial but not essential. A recruit must complete several weeks of training in animal law, animal breeds, behavior, and illness and diseases. Training also includes capture and handling procedures and instruction on how to deal with the public.

Personal Qualifications

Some of the characteristics required of an Animal Control Officer include: good communication skills, a sense of fairness, good judgment, and a genuine love of animals, especially horses. The Animal Control Officer should be physically fit in order to safely handle animals, and should be persuasive and gentle. For the most part hours are irregular, and duty requires both indoor and outdoor work. Outdoor work is performed in all types of weather conditions. Animal Control Officers work under a tremendous amount of stress because of the verbal and physical abuse they sometimes receive from citizens. They are also at risk for depression because of their exposure to animal euthanasia.

Licenses and Certification

In some states, applicants must demonstrate their proficiency in animal law enforcement and be legally certified by superiors as part of their on-the-job training. Certification and training may be obtained through the Law Enforcement Training Institute at the University of Missouri. Upon successful completion of a written examination (receipt of a score of 80 percent or above), participants are awarded certificates of completion and continuing education unit (CEUs) by the Law Enforcement Training Institute, School of Law, Extension Division, University of Missouri at Columbia. Participants who have successfully completed Level I and Level II will, upon completion, be eligible to receive their national certification.

Employment Outlook

Animal control is considered to be one of the fastest-growing professions in the United States. This profession involves public health, safety, and law enforcement. Smaller communities are usually limited in positions and advancement. Small cities sometimes operate with one- or two-person departments. Mid-sized and larger cities offer the most variety and opportunities for advancement within the animal control structure. Most local and national humane societies hire a small number of officers for field work. Hiring is usually based on the amount of funds available within a limited budget. It is not unusual to find Animal Control Officers working on a part-time basis.

Related Occupations

Occupations related to the Animal Control Officer include: Mounted Park Ranger and Mounted Police Officer.

ANNOUNCER

Description

An Announcer may be employed in various capacities within the horse industry. The following are examples of employment available to the professional Announcer:

◆ Horse Show Announcer
◆ Rodeo Announcer
◆ Racetrack Announcer
◆ Polo Event Announcer
◆ Television and Radio Announcer

Horse Show Announcer. This Announcer provides information to exhibitors, spectators, and show management. Horse Show Announcers may play music, announce classes, and announce class results, substitutions, and times for events. They also inform spectators, exhibitors, and management of emergency matters.

Rodeo Announcer. Rodeo Announcers provide a play-by-play account of each event as well as the results and times for each event. They also give the spectators a complete background on each of the contestants and the specific rodeo event, and they announce emergency information to spectators.

Racetrack Announcer. Racetrack Announcers provide information on each horse in a race to spectators at the racetrack. They also give the spectators a running description of each race as well as the final results. They are required to announce the starting time of each race and to make general announcements throughout the day.

Polo Event Announcer. Polo Event Announcers provide a play-by-play account of the match to the spectators. They also announce the beginning and end of each period of play (Chukka), announce the backgrounds of each player, and make general announcements throughout the day.

Television and Radio Announcer. The Television and Radio Announcer must be able to conduct interviews with exhibitors, management, and spectators at a particular horse event. These Announcers must convey to the audience areas of interest such as background information

on the event, exhibitors, contestants, and the horses involved. They may also be required to announce a play-by-play account of the horse event to the viewers and listeners.

Educational and Training Requirements

A high school education is essential as well as a two- or four-year college degree in Communications and Broadcasting. A general background in the area of specialization is beneficial. Additional training in electronics and public speaking is also an asset to the professional Announcer.

Personal Qualifications

The Announcer must be willing to work long hours, including weekends. Most work is performed in an enclosed area; however, Announcers may be required to work outdoors in various types of weather conditions. They must be well-organized, be pleasant, have a sense of humor, and be able to speak in a clear, loud voice. Announcers must have good communication skills, especially when dealing with the public, event managers, and contestants.

Licenses and Certification

There is no mandatory licensing or certification requirement in order to be employed as an Announcer.

Employment Outlook

The employment outlook for the Announcer in the horse industry is good. With the present increase in the horse population and horse-related events, the professional Announcer is in great demand. Now that telecommunications are becoming a vital part of equine events, the demand for the professional Announcer is on the rise.

Related Occupations

Occupations related to the Announcer include: Equine Journalist and Equine Sportscaster.

APPRAISER

Description

Appraisers are responsible for appraising horses in order to ascertain their values for various reasons, such as loan approval, issuance of insurance policies, divorce settlements, bankruptcy proceedings, disposition of estates, and horse sale auctions. They examine the horse's conformation pedigree and athletic and breeding abilities, and they determine the horse's personality and temperament. Appraisers must be aware of the current market values and economic trends within the horse industry. They prepare and submit written reports of estimated values to clients requiring their services. An Appraiser may become specialized in a particular area of the horse industry such as thoroughbred racing. (See Figure 10.2.)

Educational and Training Requirements

A high school education is a requirement for this position. A two-year degree in Equine Science with courses in business, computer science, and accounting is beneficial. Practical experience with horses as well as a thorough knowledge of anatomy and physiology, pedigrees, and breeding are also important.

Personal Qualifications

Appraisers must be willing to travel to various locations in order to perform their appraisal duties. They must have good communication skills when dealing with clients. They must be fair when establishing a value on a particular horse. Their hours are for the most part irregular, and the work is divided between physical examination of the horse and the preparation of the final written report. Work is also divided between indoors and outdoors. Appraisers should have good research skills and sound judgment when evaluating and reporting on horses and must be discrete about their findings in order to protect the privacy of their clients.

FIGURE 10.2
The appraiser is responsible for appraising horses in order to establish their value.

Licenses and Certification

No mandatory license required for the profession of an Appraiser. Certification is possible through professional organizations such as the American Society of Equine Appraisers.

Employment Outlook

Opportunities should generally be good for persons who have completed formal Equine Science programs as well as business training programs. With an increase in horse population and ownership, there will be an increase in the need for the services of the professional horse Appraiser.

Related Occupations

Occupations related to the Appraisers include: Bloodstock Agent, Breeding Technician, and Equine Attorney, and Horse Insurance Agent or Broker.

ARCHITECT

Description

Architects specializing in the horse industry design horse buildings to house horses. They also design other horse-related structures such as

turnout sheds, tack rooms, and feed and hay
buildings. They design structures to be functional, economical, as well as safe for both people
and horses. They may be involved in all phases
of development from the planning stages to the
completed structure. The Architect's duties are
numerous and skills include designing, engineering, supervisory abilities, management, and communication. Architects often work with other
professionals such as landscape architects, interi-
or designers and engineers.

Educational and Training Requirements

There are several types of professional degrees
in Architecture. Most architecture degrees are
five-year Bachelor of Architecture programs
designed for students entering from high school
or with no previous architectural training. Some
schools offer a two-year Master of Architecture
program for students with a preprofessional
undergraduate degree in Architecture or a
related area. There is also a three- to four-year
Master of Architecture Degree for students with
a degree in another area. There are many com-
binations and variations of these degree programs. A typical Architect program consists of
courses in architectural history and theory,
building design, mathematics, physical sciences,
and technical and legal aspects of architecture
and liberal arts.

Personal Qualifications

Architects must be able to communicate their
ideas to their clients; they must have good communication skills, both written and oral. They
must also have creative qualities and artistic and
drawing abilities. The Architect must be able to
work independently or as part of a team.
Computer literacy is an essential skill for the
Architect, who needs to have a thorough knowledge of CADD (computer-aided design and
drafting).

Licenses and Certification

All Architects must be licensed by the state in
which they reside and conduct business. Before a
license is issued, the candidate must meet the following requirements:

1. a professional degree in Architecture from a
   school of architecture with programs accredited by the National Architectural
   Accrediting Board
2. an internship or practical training in the
   field of Architecture for a three-year period
3. passage of all sections of the Architect
   Registration Examination

Employment Outlook

The employment of Architects depends a great
deal on the condition of the nation's economy.
The present increased trend in the horse population in the United States will translate into
the building of more horse-related facilities, and
renovation and rehabilitation of existing buildings is another area of employment for the
Architect in the horse industry. The need to
replace Architects who retire or leave the labor
force will also provide many additional job
openings in this profession.

Related Occupations

Occupations related to the Architect specializing
in the horse industry include: Equine Advertising
Specialist, Equine Real Estate Broker and Agent,
and Groundskeeper.

ASSISTANT HORSE TRAINER

Description

The duty of the Assistant Horse Trainer is to
ensure that racehorses are exercised properly and
are kept in top physical condition as per the
directions of the horse trainer. The Assistant
Trainer is responsible for the daily operation of

the racing stable by overseeing the stable employees and exercise riders. Other duties include feeding, setting out equipment, administering medication, bandaging, keeping records, ordering supplies, hiring and dismissing employees, and accompanying horses to the races. In the absence of the trainer, the Assistant Trainer is in complete charge of the racing operation. Assistant Trainers may be required to enter a horse in a race or scratch a horse from a race, to contact owners and secure veterinarians for emergency situations, and to obtain the services for a horse transport service in order to ship a horse to another racetrack or farm.

Educational and Training Requirements

A high school education and a two- or four-year college degree in Equine Science or Animal Science are beneficial. Practical experience at a racetrack as a groom, hot walker, exercise rider, or harness driver would aid the applicant in securing a position as an Assistant Trainer.

Personal Qualifications

The Assistant Horse Trainer must be alert, physically fit, and willing to work long hours each day. A typical week may include 6 or 7 days. Assistant Horse Trainers must have good communication skills in order to deal with employees, owners, and superiors. They must also have an excellent speaking voice as well as good telephone manners. The ability to speak a second language, particularly Spanish, is most beneficial. The Assistant Trainer must be willing to travel on short notice, have a genuine love of horses, and be sincerely interested in the well-being of horses.

Licenses and Certification

An Assistant Trainer's license is required by the racetrack where employment takes place. Each state varies in the requirements for licensing an

Assistant Trainer, but in all cases, a trainer must proclaim an individual as his or her assistant. A fee is paid, fingerprints and photographs are taken, and a license is issued. Some states require a specific period of employment at a racetrack prior to a license being issued, and some states require the applicant to pass a written and practical test before an Assistant Trainer's license is issued.

Employment Outlook

The employment outlook for the Assistant Trainer is generally poor, since the decline in horse racing in the United States translates into a decline in the need for the services of the Assistant Horse Trainer.

Related Occupations

Occupations related to the Assistant Horse Trainer include: Exercise Rider, Groom, and Horse Trainer.

AUCTIONEER

Description

The main function of the Auctioneer is to sell horses to the highest bidder at public horse sales. Auctioneers may be self-employed or employed by a large horse sales company. Some of the basic duties of the Auctioneer include examining horses for sale and offering advice to clients on the legal aspects of a horse auction. Auctioneers determine whether sellers wish to establish reserve bids on their horses. A reserve bid is the lowest figure a seller will accept for a horse at an auction. Auctioneers arrange for the location of auctions, set up all advertising concerning auctions, and prepare sales catalogs listing all the horses for sale as well as their pedigree, racing, and show record. Throughout a sale, Auctioneers set the pace for both buyers and sellers. They constantly point out the good features of the horses in the sales ring and, through their chant, obtain the best prices possible for the horses being sold.

Educational and Training Requirements

A high school education is essential as well as vocational training in Equine Science. College courses in business, accounting, economics, and marketing are beneficial. Technical training at a school of auctioneering and practical experience as a spotter, cashier, and horse handler at a large horse auction firm are essential.

Personal Qualifications

Auctioneers work in all types of weather conditions. They work long hours and under a great deal of pressure, and they do a great deal of traveling throughout the year. They must have good communication skills and enjoy working with people. The Auctioneer must have a good speaking voice, a working knowledge of electronic public address systems, and a good sense of humor in order to be entertaining during the sale proceedings. Finally, auctioneers should be knowledgeable in conformation, veterinary medicine, and equine market trends.

Licenses and Certification

All states have laws to regulate livestock auction sales. In some states, the Auctioneer is required to pass an examination and pay a mandatory licensing fee. Other states require the Auctioneer to acquire a stated number of hours of auctioneering schooling. Certification can be obtained through professional organizations such as the National Auctioneers Association or the Certified Auctioneers Institute.

Employment Outlook

There is a great demand for skilled Auctioneers. Auctions are one of the most popular methods of selling horses. Horse associations and horse ownership have increased the number of horse and pony auctions throughout the country. The growing popularity and success of horse auctions should create a demand for more professional auctioneers.

Related Occupations

Occupations related to the Auctioneer include: Appraiser, Bloodstock Agent, and Equine Accountant or Bookkeeper.

AUCTION SALES EMPLOYEE

Description

In order to be successful, a horse auction sales company must have a large staff ranging from the auctioneer to the person holding the horse in the sales ring during bidding. Before the actual auction begins, the **Catalog Staff** must put together a complete catalog of all the horses consigned to the sale. They must compile information on each horse such as racing record, show record, complete pedigree, monies and awards earned, and breeding records.

The **Auctioneer** conducts the bidding process by chanting the bids aloud to the buyers in the audience. The **Spotter** is responsible for passing the bid from the prospective buyer in the audience to the Auctioneer. The **Sales Clerks** are responsible for the collection of all funds from entry fees to final sales price. They must communicate all telephone bids with absentee buyers during the sale. They are also responsible for the registration papers of all horses in the sale and for all documents relating to the transfer of ownership to the new owner. Finally, they tabulate the results of the sale through statistics indicating information such as the highest price paid, average price per horse, leading consignors, and leading buyers. The **Horse Handler** actually holds and controls the horse while it is in the sales ring during bidding.

Educational and Training Requirements

The basic educational requirement consists of a high school diploma. A two- or four-year college degree in Equine Science or Business Management is beneficial. College-level courses should include marketing, economics, and computer science. Naturally any practical experience with horses or knowledge of horsemanship is beneficial for the Auction Sales Employee.

Personal Qualifications

The Auction Sales Employee must be willing to work irregular hours, including weekends. Most sales work is performed indoors in an office setting. Occasionally work is performed outdoors, for example, inspecting horses for select sales or conducting a sale at a farm location. Travel is considered an essential of this profession.

Licenses and Certification

There is no mandatory license or certification requirement for the horse Auction Sales Employee.

Employment Outlook

Horse sales provide an excellent outlet for marketing horses, regardless of the breed. With a growing increase in horse ownership, the future of horse auction sales is secure. This will result in a growing need for the horse Auction Sales Employee.

Related Occupations

Occupations related to the Auction Sales Employee include: Auctioneer, Bloodstock Agent, Groom, and Pedigree Researcher.

BLOODSTOCK AGENT

Description

The duty of the Bloodstock Agent is to buy and sell horses for clients. Bloodstock Agents may be employed by a bloodstock agency, or they may be self-employed. Bloodstock Agents perform most of their duties at public horse sales. Their main objectives are to obtain the best quality of horse for their clients at the lowest possible price and to sell their clients' horse at the highest possible

price. They must research the market trends and pedigrees, follow up on past purchases, and be able to evaluate a horse's potential for the purpose for which it will be used. The thoroughbred industry provides an excellent source of income for the Bloodstock Agent, more so than any other breed of horse because of the thoroughbred's high market value. Bloodstock Agents may also provide other services to their clients in addition to buying and selling horses. These services may include: insurance, transportation, syndications, appraisals, accounting, advertising, sales preparation, boarding, and breaking and training. Bloodstock Agents receive monetary compensation for their services in the form of a commission on the horses they buy and sell for their clients.

Educational and Training Requirements

A high school education is essential as well as a two- or four-year college degree in Equine Science or Animal Science. College courses in business, accounting, economics, and marketing are beneficial. Practical experience with horses as an employee of a breeding farm, sales company, or bloodstock agency is helpful.

Personal Qualifications

Bloodstock Agents should enjoy working with both horses and people. They must have good communication skills, both oral and written, when dealing with people. They must have an excellent eye for conformation when selecting horses for clients. Bloodstock Agents must be able to make quick decisions and must be resourceful, confident, and patient. They are required to work outdoors in all types of weather conditions, and they work long hours and under a great deal of pressure. They must be willing to travel both nationally and internationally to attend selected horse sales.

Licenses and Certification

There is no mandatory licensing or certification requirement for the occupation of Bloodstock

Agent. However, there are several organizations that offer membership to the Bloodstock Agent, such as the Livestock Marketing Association and the Independent Livestock Marketing Association.

Employment Outlook

The number of positions for Bloodstock Agents is limited because it is essentially a self-employed occupation. Competition in this field is high, and beginners are advised to gain employment within the horse industry to gain experience and to be available when a Bloodstock Agent position develops.

Related Occupations

Occupations related to the Bloodstock Agent include: Appraiser, Auctioneer, Equine Accountant or Bookkeeper, Horse Insurance Agent or Broker, Horse Trainer, and Pedigree Researcher.

BREEDING TECHNICIAN

Description

There are two basic types of Breeding Technicians:

1. Artificial Insemination Technician
2. Live Breeding Technician

The reason for these two types of Breeding Technicians is that some horse breed associations permit artificial insemination, but others allow only live breeding (that is, actual physical contact between the stallion and mare).

Artificial Insemination Technician. The duty of the Artificial Insemination Technician is to deposit prepared stallion semen into the uterus of the mare in order to impregnate the mare. Artificial insemination allows the use of outstanding stallions, especially those that have produced offspring that exhibit beneficial genetic traits designed to improve the breed. Through artificial insemination the breeder can extend the use of

the stallion. One stallion can sire a large number of offspring in one season. Artificial insemination also aids in the control of diseases, as there is no physical contact between the stallion and the mare. The Artificial Insemination Technician is responsible for:

◆ collection and evaluation of semen from the stallion

◆ storage and preparation of semen

◆ transporting of semen

◆ inseminating and diagnosing pregnancy of the mare

◆ teasing and rectal palpation of the mare

Live Breeding Technician. This type of breeding technician is part of the breeding staff within the breeding shed when a live cover by a stallion is required. This would be the case with the thoroughbred breed. When breeding a thoroughbred stallion to a thoroughbred mare, several Live Breeding Technicians must be on hand. They are responsible for:

◆ controlling the stallion and mare during live breeding

◆ washing the reproductive organs of the stallion and mare before and after breeding

◆ applying restraining and protective equipment to the stallion and mare

◆ physically "teasing" the mare prior to breeding to determine the estrus period

◆ maintaining complete records on all live matings

◆ cleaning and disinfecting the breeding shed and all equipment utilized during live breeding

Educational and Training Requirements

A high school education is required by most employers. A horse background is beneficial for this position. A two-year degree or a certificate from a private technical school or an organization such as the National Association of Animal Breeders (NAAB) is desirable but not a requirement. Certain areas of study, which might include agriculture, animal science, economics, genetics, and livestock reproduction, are valuable for obtaining key positions within the horse breeding industry. In addition to a formal education, on-the-job training is most beneficial for this position.

Personal Qualifications

Breeding Technicians must be able to work independently with little supervision. They must enjoy working with both people and horses. In order to perform their duties, they must be able to communicate well. They should possess good business skills. Both types of Breeding Technicians perform their duties indoors as well as outdoors in all types of weather conditions. They must be healthy and physically fit in order to perform their daily duties. During the active breeding season, a Breeding Technician may work as much as 12 hours per day depending on the number of mares to be bred. Most Breeding Technicians will work a 5- or 6-day week, depending on their workload.

Licenses or Certification

Some states require each technician to earn a certificate in the study of Artificial Insemination from an accredited educational facility. The NAAB does not certify or license Artificial Insemination Technicians, but a certificate of completion is usually granted to those technicians completing the course. Note that the NAAB is focused on the cattle industry; only about one member out of 24 is involved with artificially inseminating horses.

Employment Outlook

Breeding horses by means of artificial insemination is accepted and permitted by certain horse breed associations. For example, thoroughbreds may not be bred by means of artificial insemination under any conditions by ruling of the Jockey Club. The Jockey Club is responsible for the registration of all thoroughbreds born in the United States. Most horse breeders depend on artificial insemination to obtain the greatest breeding efficiency from their stallions. This will result in a demand for the professional Artificial Insemination Technician in the near future. The Live Breeding Technician is in demand because of the increase in the number of thoroughbred stallions now standing stud in the United States. There is an increasing trend within the thoroughbred breeding industry to ship stallions to countries located in the southern hemisphere during the summer months in order to obtain a year-round income from breeding stallions.

Related Occupations

Occupations related to the Breeding Technician include: Equine Veterinary Assistant, Farm or Ranch Manager, Groom, and Veterinarian.

CLOCKER

Description

The responsibility of the Clocker is to time horses to determine how fast they are traveling for a specific distance as they work out during the morning exercises. Clockers are responsible for recording the time for each racehorse that is working or "breezing." As the racehorse enters the racetrack in the morning, the rider or trainer of the horse must state to the gap attendant the name of the horse, the distance it is going to work, and the name of the trainer. The gap attendant will then relay this information to the official Clockers situated in the clocker's booth overlooking the racetrack. After the workout, the Clocker will report the official time back to the gap attendant, who in turn will advise the rider or the trainer of the official time. After the morning workouts are over, the official times are reported to the various racing publications and the racing authorities. The times are then published in racing journals and official programs for public information. Clockers also add their comments concerning a workout, such as "from the starting gate, handily, driving."

Educational and Training Requirements

No formal education is required for this position other than a basic high school diploma. A good background in horsemanship skills is beneficial as well as some practical experience on the backstretch as a groom, exercise rider, or stablehand. Finally, clockers must be able to operate and read a stopwatch properly.

Personal Qualifications

Clockers must be at their post early in the morning before sunrise. They are subject to all types of weather conditions, as racing occurs year-round in all parts of the country. They must have good eyesight and be able to focus on several horses working out at the same time. They must be familiar with the various stables on the grounds and their riders so as to time the correct horse at the correct distance. The Clocker must have good communication skills and a pleasant attitude when dealing with trainers, journalists, and racetrack officials.

Licenses and Certification

There is no specific license or certification required for a Clocker. The only requirements are those that apply to all racetrack employees, which entail fingerprinting, photographs, and background checks.

Employment Outlook

Clockers are hired at all racetracks offering flat racing and harness racing. Some Clockers are employed by official racing publications such as the *Daily Racing Form.* As with any racetrack position, the present decline in the number of racetracks will result in a decline in the demand for the services of professional Clockers.

Related Occupations

Occupations related to the Clocker include: Assistant Horse Trainer, Exercise Rider, Groom, and Horse Trainer.

EQUINE ACCOUNTANT OR BOOKKEEPER

Description

The Equine Accountant or Bookkeeper keeps track of finances, keeps records, and prepares and aids in paying taxes for all equine-related operations. Equine Accountants or Bookkeepers provide equine-related businesses with up-to-date figures on profit and loss, depreciation, wages and taxes, debts and assets, and other financial data. They must record, sort, and verify figures in order to keep track of all income and expenses for their clients. Bookkeeping clerks compile figures and expenses, accounts payable, accounts receivable, and profit and loss information. They prepare financial reports for management. Accounting clerks post and update all transactions of the equine-related business into computer files. They analyze computer printouts with actual written ledgers and make corrections if needed. Both bookkeeping and accounting clerks may be responsible for bank deposits, audit preparation, mailings, and filing.

Educational and Training Requirements

A high school education is required for this position. High school courses should include computer literacy, bookkeeping, business mathematics, and typing or keyboarding. In addition to a high school education, a two- or four-year college degree in Accounting or Business Management is also essential for this position. Practical work experience in the horse industry is a strong asset for the beginner as well.

Personal Qualifications

Accounting and bookkeeping clerks must be detail-oriented, accurate, and orderly. They must be neat and enjoy working with people and figures. They must be discreet and trustworthy. Most of their work is performed indoors in an office setting. They must be able to work well with others and possess good communication skills. They must have a keen knowledge of the specific equine business in which they are involved. Finally, accounting and bookkeeping clerks must be willing to work overtime during tax periods, inventories, audits, and at the end of each year.

Licenses and Certification

Some states require all accounting and bookkeeping clerks who deal in tax returns and tax-related matters to be licensed. In order to become a Certified Public Accountant (CPA), one must pass a required written examination in the state in which one resides and conducts business.

Employment Outlook

The volume of horse-related business transactions is expected to grow in the near future; however, the increased productivity of accountants and bookkeepers will reduce the rate of employment opportunities. The main reason for this increase in productivity is automation and the widespread use of the computer. Employment opportunities will also be reduced by the centralization of many office operations, although new job openings will occur as accountants and bookkeepers move on to other occupations or retire.

Related Occupations

Occupations related to the Equine Accountant or Bookkeeper include: Bloodstock Agent, Equine Attorney, and Equine Real Estate Broker and Agent, and Horse Insurance Agent or Broker.

EQUINE ADVERTISING SPECIALIST

Description

The duty of the Equine Advertising Specialist is to oversee the marketing policy of a horse-related business. This includes market research, strategy, sales promotions, advertising, and public relations. Equine Advertising Specialists determine the demand for horses and services offered by the farm and its competitors. They determine the best methods to promote the farm's horses and services to attract clients, and they analyze horse sales statistics in order to determine the sales potential of the horses they are marketing as well as the breeding preferences of clients. They also monitor economical and political trends that might affect the horse industry as a whole and make recommendations to clients on how to meet these trends. Their clients include breeding farms, racing and show stables, horse sales companies, racetracks, feed companies, veterinary pharmaceutical companies, and horse equipment manufacturers.

Educational and Training Requirements

A high school education is mandatory for this position. A two- or four-year college degree in Liberal Arts or Business Administration is essential. College courses should include psychology, philosophy, sociology, economics, accounting, business law, and creative and technical writing. Knowledge of horses is not required but is beneficial to Equine Advertising Specialists for understanding the needs of their clients.

Personal Qualifications

Equine Advertising Specialists must have excellent communication skills when dealing with employers, clients, and fellow workers. They must be creative mature, and express good judgment. They must be willing to work long hours, including evenings and weekends, and they must be able to handle stress when confronted with deadlines and standards. Travel is an essential element of this profession, as it is a requirement when meeting with clients and media representatives. Computer literacy and word processing are definite requirements for this profession. Most work is performed indoors in an office setting at a desk or layout table.

Licenses and Certification

There is no specific license requirement for the profession of Equine Advertising Specialist.

Some associations, such as the American Marketing Association and the American Advertising Federation, offer certification programs in the field of advertising. Certification is a symbol of confidence and achievement in the field of marketing and advertising.

Employment Outlook

The employment outlook for the Equine Advertising Specialist is good. Most people in the horse industry are beginning to realize that horses are a business and that they must adhere to modern business practices in order to survive economically. Intense competition within the horse industry, both domestic and international, is creating a demand for the professional Equine Advertising Specialist.

Related Occupations

Occupations related to the Equine Advertising Specialist include: Equine Accountant or Bookkeeper, Equine Artist, and Equine Journalist.

EQUINE ARTIST

Description

The Equine Artist utilizes a variety of methods and materials to capture the essence of the horse. Equine Artists can generally be divided into two categories:

1. Graphic Artist
2. Fine Artist

Graphic Artists, who usually have their own studios, put their vision and artistic skills involving the horse to the service of commercial clients such as corporations, retail stores, and advertising and publishing firms. Graphic Artists create packaging, promotional displays, and marketing brochures for new horse-related products. They are responsible for the layout design of horse industry publications, and they create the horse-related graphics for television and other types of media.

Fine Artists, on the other hand, create artistic work involving the horse to suit their own need for self-expression. They usually display their work, which emphasizes the equine subject, in galleries, private homes, and museums. Fine Artists usually work independently and utilize their preferred medium; they become specialists in the field as painters, sculptors, illustrators, cartoonists, photographers, animators, or art directors. Fine Artists also work on commission for clients such as portrait painters, sculptors, and photographers.

Educational and Training Requirements

A high school education plus a two- or four-year degree in Fine Arts is essential. Formal training requirements are virtually nonexistent, but some basic training is essential to develop skill as an artist. Knowledge of equine behavior, anatomy, physiology, and general conformation is essential; and a general knowledge of equine equipment, supplies, buildings, and landscapes is beneficial. As in any self-employment situation, a knowledge of business and computer skills is helpful, along with a knowledge of art history. The Equine Artist must put together a professional portfolio to show to clients.

Personal Qualifications

Equine Artists must have natural artistic ability and be creative and imaginative. They must have excellent communication skills when dealing with clients and agents, and they must be able to work independently, as most artists are self-employed and possess a general passion for the equine subject. Most work is performed indoors; however, some work may be performed outdoors on location. The Equine Artist must be willing to travel in order to work with equine subjects. Good photographic skills may be necessary in order to capture a particular equine subject for color, lighting, and conformation detail. Finally, Equine Artists must have a passion for the work and a personal drive, as rejection is part of the process of exhibiting and selling their work.

Licenses and Certification

There is no mandatory licensing or certification requirement for the Equine Artist.

Employment Outlook

The number of aspiring artists will continue to exceed the number of available positions, which will result in extreme competition for salaried positions as well as freelance work. Fine Artists will find it very difficult to earn a living solely by selling their artwork until they obtain the necessary experience and establish a good relationship with the art world or the horse industry. Fame and fortune do not necessarily go hand in hand; there are constant expenses with little return.

Related Occupations

Occupations related to the Equine Artist include: Equine Advertising Specialist, Equine Photographer, and Equine Videographer.

EQUINE ATTORNEY

Description

Attorneys who practice Equine Law represent one of the opposing parties in equine-related civil or criminal trials by presenting evidence that supports their clients in court. They also act in an advisory capacity in equine-related matters by counseling their clients regarding legal rights and obligations, as well as providing advice in all business and personal matters. Attorneys aid their clients in computing taxes for their equine business and explore various tax strategies for their clients. They advise what actions their clients may pursue, and they prepare legal documents such as wills, partnerships, and syndications. Attorneys also create contracts pertaining to the horse industry, such as contracts involving breeding, boarding, insurance, sales, and transportation.

Educational and Training Requirements

An Attorney must complete at least three years of college and graduate from a law school approved by the American Bar Association (ABA) or by the proper state authorities. The required college and law school education usually takes seven years of full-time study after high school—four years of undergraduate study followed by three years of law school. Graduates receive the degree of Juris Doctor (JD) or Bachelor of Law (LL.B.) as the first professional degree.

Personal Qualifications

Attorneys specializing in Equine Law require proficiency in the skills of writing, reading and analyzing, thinking logically, and communicating verbally. They should like to work with people and be able to gain the confidence and respect of their clients, associates, and the public. Honesty and integrity are required characteristics for this occupation. Attorneys must be creative while working on new and unique legal problems. Finally, perseverance and reasoning ability are essential traits when dealing with complex cases. Attorneys must be willing to work long, irregular hours.

Licenses and Certification

To practice law in the courts of any state, a candidate must be licensed or admitted to the state bar under the rules established by the jurisdiction's highest court. Applicants for admission to the bar must pass a written bar examination. Some states require applicants to pass a separate written ethics examination.

Employment Outlook

As the horse industry has become more complex with contracts, taxes, and insurance and business issues, the professional Equine Attorney has become a necessity. A growing demand for Attorneys will result directly from the growth in the horse population and general equine business activities. Competition for job openings, however, will continue to be high because of the large number of law school graduates each year.

Related Occupations

Occupations related to the Equine Attorney include: Equine Journalist and Mounted Police Officer.

EQUINE BOOK DEALER

Description

The function of the Equine Book Dealer is to buy and sell horse-related books and videos to the general public. Some Equine Book Dealers distribute new books and videos, while others specialize in rare and scarce books on horses or in one discipline, such as dressage, racing, showing, or breeding. These Book Dealers must be knowledgeable in their specific area of expertise. Equine Book Dealers must be able to assist customers in searching for a specific book or video, maintain a complete inventory of books and videos, and set and implement all policies, goals, and procedures concerning the overall operation of the bookstore. They also must supervise employees in pricing books and videos, setting up displays within the store, and organizing shelves. Those who specialize in rare and scarce books must attend public and estate auctions in order to obtain them for their inventory.

Educational and Training Requirements

A high school education is essential. A college degree is beneficial but not mandatory. Educational courses should include equine studies, accounting, economics, computer science, and business management. Any practical work experience in a retail trade position, such as sales clerk, store manager, or customer service representative, is most beneficial. Practical horse experience is not mandatory, but a general knowledge of horsemanship is helpful in understanding the needs of customers.

Personal Qualifications

The Equine Book Dealer must have good business sense, customer service skills, and marketing abil-ity. Most work is performed indoors within a store or private residence. Travel may be required in order to attend book auctions, meet with publishers or distributors, and attend book fairs and trade shows. Equine Book Dealers must have good communication skills, both written and oral, when dealing with customers and suppliers. They are usually self-employed, which may require long, irregular hours, especially during holidays. Working weekends and evenings is considered the norm in any type of retail business.

Licenses and Certification

There is no mandatory licensing or certification requirement for the Equine Book Dealer.

Employment Outlook

With the growing trend in the purchase of books and videos from mail-order catalogs and shopping on the Internet, the employment outlook for the Equine Book Dealer is poor. The location of the bookstore and the quality of inventory is crucial in achieving a successful book dealership. As with any retail business, the condition of the overall economy plays a key role in determining the success of book dealerships.

Related Occupations

Occupations related to the Equine Book Dealer include: Equine Librarian, Feed and Grain Dealer, and Retail Tack Shop Operator.

EQUINE CHIROPRACTOR

Description

Veterinary chiropractic is the examination, diagnosis, and treatment of nonhuman animals through manipulation and adjustments of specific joints. The term *veterinary chiropractic* should not be interpreted to include dispensing medication, performing surgery, injecting medications, recommending supplements, or replacing traditional veterinary care. Equine Chiropractors diagnose and treat musculoskeltal conditions of the spinal column and extremities of the horse in order to correct abnormalities of the horse's body believed to be caused by interference with the nervous system. They examine the equine patient in order to determine the nature and the extent of the disorder, and they manipulate the spinal column and extremities to adjust, align, or correct abnormalities caused by a neuralgic dysfunction. Equine Chiropractors sometimes utilize additional measures such as exercise, rest, water, heat, ultrasound, and nutritional therapy. They strive to allow the horse to reach its full potential as an athlete and to reduce pain without the use of drugs.

Educational and Training Requirements

A high school education is required for this position, as well as a doctorate degree in Veterinary Medicine or Chiropractic Education. College-level courses should include equine anatomy and physiology, pathology, and neurology. Practical work experience within the horse industry is also a strong asset.

Personal Qualifications

Equine Chiropractors must be willing to travel in order to service their clients. They must be able to detect and diagnose physical abnormalities plaguing the equine patient, and they must have considerable hand dexterity in order to perform manipulations on the horse. They must be able to work independently and handle responsibility. In addition, they must be understanding, compassionate, and possess the desire to help the equine patient. Finally, they must have good communication skills when dealing with horse owners, veterinarians, and office staff.

Licenses and Certification

The Equine Chiropractor can receive certification through professional organizations such as the American Veterinary Chiropractic Association (AVCA), which sponsors an Options Certification Course in animal chiropractic. The purposes of the AVCA are to:

- certify educational programs in animal chiropractic provided by Options for Animals
- provide a standard of care for the practice of animal chiropractic
- promote the application of chiropractic in animal health care
- provide information about animal chiropractic care to animal owners
- encourage and sponsor research in animal chiropractic
- encourage interprofessional interaction and cooperation for the betterment of animal health

The basic certification course consists of five modules, each with 30 hours of instruction. The first four modules may be taken in any order. The last, or integrated, module must be taken after completion of the first four modules. Chiropractors who treat humans and then decide to specialize in equine chiropractic medicine must meet the educational requirements and pass a state board examination. Chiropractors may only practice in the states in which they are licensed.

Employment Outlook

The employment outlook for the Equine Chiropractor is good. As public awareness in this profession grows within the horse industry, the need for the professional Equine Chiropractor will also grow.

Related Occupations

Occupations related to the Equine Chiropractor include: Equine Massage Therapist and Veterinarian.

EQUINE CONSULTANT

Description

The duty of the Equine Consultant is to advise clients in matters pertaining to the horse industry such as racing, breeding, showing, sales, investments, insurance, legal matters, syndications, and real estate. Equine Consultants obtain research and analyze information concerning a particular horse-related project for a client. They then make recommendations to clients and assist them in implementing their recommendations. Clients sometime requires an Equine Consultant's expertise in pursuing a course of action in a particular horse-related matter. Equine consulting firms range in size from individual practitioners to large international organizations. Consulting services usually are provided to a client on a contract basis.

Educational and Training Requirements

The basic educational requirements consist of a high school diploma and a bachelor's or master's degree in Equine Studies, Computer Science, or Business Management. A degree in Veterinary Medicine is also valuable for this profession. In addition to the appropriate formal education, several years of experience in an equine specialization such as health, law, insurance, and so on are also beneficial.

Personal Qualifications

Equine Consultants should be creative in developing solutions to problems. They should be self-motivated and disciplined, and they must have good communication skills, both written and oral, good judgment, and the ability to manage time efficiently. They must be able to work with little or no supervision. Equine Consultants perform most of their work indoors in an office setting, but there are occasions when they visit with their clients at their residence or place of business. They are required to travel in order to meet with clients on a regular basis and to attend horse sales and seminars. Finally, Equine Consultants must be able to deal with stress, especially when working under deadlines.

Licenses and Certification

There is no mandatory licensing or certification for the profession of Equine Consultant.

Employment Outlook

Growth is expected for both large and small Equine Consulting firms that specialize in a specific area of the horse industry. The horse industry in the United States has developed into a multibillion dollar business in both domestic and international markets. Competition is fierce, and Equine Consultants are relied upon to reduce costs, develop efficient operations, and develop marketing strategies for their clients.

Related Occupations

Occupations related to the Equine Consultant include: Bloodstock Agent, Equine Accountant or Bookkeeper, Equine Attorney, and Horse Insurance Agent or Broker.

EQUINE DENTIST

Description

The main function of the Equine Dentist is to "float" the molar teeth of the horse. Floating refers to a procedure whereby the sharp enamel points of the molars are removed with the aid of a dental file. In addition to floating the teeth, Equine Dentists must be able to remove or pull the caps, considered the milk teeth, that are lightly attached to the gum. They may be required to perform another routine procedure that involves pulling the wolf teeth. These tiny wolf teeth tend to interfere with the bit and should be removed. The Equine Dentist is also called upon to remove abscessed and fractured teeth, since a sinus infection can result from a badly abscessed tooth. Finally, an Equine Dentist may be called upon to determine the exact age of a horse by examining the teeth. (See Figure 10.3.)

Educational and Training Requirements

A high school education is required for this profession. A two- or four-year degree in Animal Science or Veterinary Technology is also beneficial. A doctorate degree in Veterinary Medicine may be required in some states to perform equine

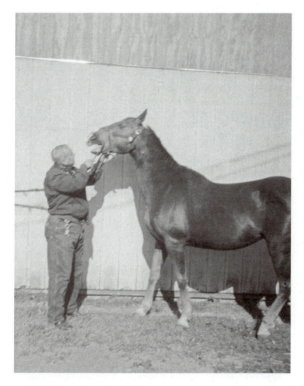

FIGURE 10.3
The Equine Dentist is responsible for maintaining the general health of the horse's teeth.

dentistry. An apprenticeship period with a licensed veterinary practitioner is beneficial in learning equine dentistry.

Personal Qualifications

Equine Dentists must be strong in order to perform hard physical work. They must have good communication skills as well as patience when dealing with both clients and their horses, and they must be willing to travel and to drive from one appointment to another. They must have a thorough knowledge of the anatomy of the horse's teeth and skull. Most work is performed indoors, inside a stall within the confines of a barn, but there are occasions when work must be performed outdoors. Most Equine Dentists work 6 days per week and an average of 8 hours per day. They usually spend 30 minutes to one hour

with each animal, depending on the amount of dental work being performed.

Licenses and Certification

Equine dentistry can only be practiced legally by a licensed veterinarian in most states, since equine dentistry is considered to be a medical procedure and the Equine Dentist may be required to administer a general anesthesia, antibiotic, or other drugs to equine patients. Minor surgery may also be performed in some cases. Some veterinarians do not like to perform routine dentistry work, as it is too time-consuming, demands hard physical labor, and simply does not command a high fee. This attitude allows the so-called Equine Dental Technician to exist. However, the Dental Technician should be trained by a licensed veterinarian and work closely with licensed veterinarians in a particular area. Certification requires the candidate to pass a written and practical examination issued by the board of directors or certification committee of the various professional equine dental organizations. Information and certification in the field of equine dentistry may be obtained by contacting professional organizations such as the International Association of Dental Technicians, the American Association of Equine Practitioners, the American Veterinary Medical Association, and the North American Veterinary Technician Association.

Employment Outlook

The employment outlook for the Equine Dentist is expected to grow in the near future. With an increase in horse ownership, the need for the services of the Equine Dentist will also increase. However, until the matter of licensing is resolved, Equine Dentistry will be limited to licensed veterinarians only.

Related Occupations

Occupations related to the Equine Dentist include: Equine Veterinary Assistant, Laboratory Animal Technician, and Veterinarian.

EQUINE EDUCATOR

Description

Equine Educators may be employed at the high school level as part of a vocational agricultural program, or they may be employed as college professors or associate professors for an Equine Science program. The duty of the Equine Educator is to teach Equine Science to students in public or private schools. Equine Educators organize the horse-related curriculum with practical and technical instruction, including demonstrations of those skills required in the field, and provide lectures on techniques, theory, and vocabulary. They organize and evaluate the work performed by students as individuals or in small groups, and they prepare tests on the subject matter to evaluate the achievement of students in general knowledge and trade skills. They may also be involved in job placement and in coordinating on-the-job training for students. Equine Educators sometimes prepare or assist in preparing budgets for their programs. Professors and associate professors at colleges conduct Equine Science courses for undergraduate or graduate students and may be required to teach one or more subjects, such as equine nutrition, genetics, and anatomy, within the prescribed curriculum. They may be involved in supervising research conducted by other staff members or graduate students working toward an advanced academic degree, and they may act as advisors to students in both academic and vocational areas. Their duties may also include acting in an advisory capacity to student horse-related organizations and clubs. (See Figure 10.4.)

Educational and Training Requirements

In order to teach on a high school level, one must have a high school education plus an associate degree, bachelor of science degree, or master's degree in the fields of Animal Husbandry, Agriculture, or Education. A master's or doctorate degree is mandatory to teach on a college or university level in the fields of Animal Husbandry, Genetics, Equine Psychology, Equine

FIGURE 10.4
The duty of the Equine Educator is to teach Equine Science to students enrolled in private and public schools.

Nutrition, and related fields of study. Consistent with teaching requirements in any field, one must serve as a student teacher sometime during one's educational training.

Personal Qualifications

The Equine Educator must be a well-organized person. Hours are for the most part irregular. Work is performed in both classrooms and outdoor facilities. Equine Educators must have an excellent speaking voice as well as excellent writing skills, and they must enjoy working with horses as well as people. Computer literacy is essential for any teaching position.

Licenses and Certification

All states require licensing and certification of teachers. Licensing and certification by the Department of Education is required for the elementary and high school teachers by the state in which the educator is employed. All applicants are required to have completed specific courses in education and successfully completed an internship teaching program.

Employment Outlook

The employment outlook for the Equine Educator is considered to be good, since an increasing number of educational institutions are offering

Equine Studies programs at both the high school and the college level.

Related Occupations

Occupations related to the Equine Educator include: Equine Extension Service Agent, Equine Librarian, and Veterinarian.

EQUINE EXTENSION SERVICE AGENT

Description

The duty of the Equine Extension Service Agent is to conduct and organize an equine cooperative extension program to aid horse owners and individuals within the horse industry in the application of equine research. This research is usually conducted primarily by land-grant colleges throughout the United States. Equine Extension Service Agents collect, analyze, and evaluate research data, and assist horse owners and individuals in solving horse-related problems. They may also deliver lectures and write articles addressing equine subjects such as nutrition, pasture management, and health issues. They prepare activities and other reports, maintain program records, prepare budgets for all activities, and supervise and direct 4-H and FFA club activities. Equine Extension Service Agents may be required to supervise and coordinate the activities of other county extension employees. Finally, they may deliver lectures to commercial and community horse organizations.

Educational and Training Requirements

A high school education as well as vocational training in stable management are essential for this profession. A college degree in Equine Science, Animal Science, or Agriculture is also beneficial. Courses in education and business management will aid the applicant in securing the position of Equine Extension Service Agent.

Personal Qualifications

Equine Extension Service Agents must have a pleasant personality, the ability to speak in public, and excellent writing skills. The use of a personal computer is essential. They must be willing to travel throughout their territory to attend seminars, give lectures, and participate in community horse-related activities. They are required to work both outdoors and indoors. When working outdoors, they are subjected to all types of weather conditions. Indoor work is usually conducted in an office setting as well as in various types of farm buildings. Because of the large amount of travel required, they must be willing to drive for long periods of time.

Licenses and Certification

There is no specific licensing requirement for the Equine Extension Service Agent. In some states, a civil service examination is required.

Employment Outlook

The employment outlook for the Equine Extension Service Agent is good but limited to the number of land-grant colleges. Presently, there are 49 land-grant colleges in the United States.

Related Occupations

Occupations related to the Equine Extension Service Agent include: Equine Educator and Equine Librarian.

EQUINE FENCE DEALER OR INSTALLER

Description

Fences are used on horse farms and ranches to enclose paddocks, driveways, barns, homes, and pastures. The main function of any type of fence located on horse property is to confine the horses, distinguish property boundaries, provide privacy, and discourage trespassers. Equine Fence Dealers or Installers sell, install, and service various types of fences used on horse farms and ranches. Some represent one fence manufacturing company and sell and install only that company's fences, while others are independent dealers who sell and install different fence manufacturers' products. These dealers work directly with customers, manufacturers, and installers. Most fence dealerships are located in rural areas. The main responsibility of the Fence Installer is to install and repair fences and gates that are made of wood, vinyl, plastic, chain link, and aluminum. Duties include laying out the fence line, digging post holes with a post-hole digger or a power-driven auger, then setting the upright posts into a concrete mix. Once the upright posts are installed and leveled, the Installer attaches the fence railings. The Fence Installer may be required to use a power saw on wooden fences and a portable welding unit on metal fences.

Educational and Training Requirements

A high school education is required for the Fence Dealer or Installer. High school courses should include mathematics, English, wood shop, metal shop, welding, and drafting. A Fence Dealer should also have a two- or four-year degree in Agriculture, Animal Science, or Agribusiness. A horse background is beneficial for both positions.

Personal Qualifications

The Equine Fence Dealer or Installer must be responsible, dependable, and should enjoy working with people. The Fencer Dealer must have good communication skills when dealing with clients, manufacturers, and installers. The Fence Installer must be able to perform physical tasks and work with various types of tools. Most of the Installer's work is performed outdoors in all types of weather conditions.

Licenses and Certification

There is no specific license or certification required for the positions of Equine Fence Dealer or Fence Installer. Installers may become a member of a labor union such as the Laborer's International Union of North America (AFL-CIO).

Employment Outlook

For the most part, employment opportunities are good for the Equine Fence Dealer or Installer, since boundary lines, privacy, and security are in demand on horse farms and ranches, and the demand will continue in the near future.

Related Occupations

Occupations related to the Equine Fence Dealer or Installer include: Equine Architects, Farm Equipment Dealer, and Groundskeeper.

EQUINE GENETICIST

Description

Equine Geneticists are concerned with genetic heredity in horses. They study the process by which individual characteristics are transmitted. They study the role of genes in the development of the individual and the methods of producing new traits that will enhance a particular breed of horse. The Equine Geneticist is also concerned with the effect of the environment on inherited characteristics. Most Equine Geneticists are involved in either research or teaching. Those involved in research are usually employed by a college, university, veterinary medical school, or veterinary pharmaceutical company. Research projects may focus on the development of new strains and varieties of horses, the transmission of inherited equine diseases, and the concept of genetic segregation. Equine Geneticists prepare and conduct laboratory studies and prepare reports on their research. Those who are involved in teaching at colleges and universities conduct classes, supervise students in laboratory work, prepare lessons, serve on professional committees, write articles, give lectures, and prepare papers.

Educational and Training Requirements

A high school education with an emphasis on mathematics, biological sciences, physical sciences, and language arts is essential for this profession. A bachelor of science degree in Chemistry, Physics, or Mathematics is essential. Finally, a master's or doctorate degree in Genetics is required for this profession.

Personal Qualifications

Equine Geneticists must have a genuine interest in science and the equine species. They should have good communication skills when dealing with students, fellow scientists, technicians, and laboratory personnel. Other personal characteristics of the Equine Geneticist include good judgment, patience, perseverance, integrity, and initiative.

Licenses and Certification

There is no specific licensing requirement for the profession of Equine Geneticist.

Employment Outlook

The employment outlook for the Equine Geneticist is expected to expand in the near future, especially in areas of medical and clinical genetics and genetic engineering. Teaching positions for the Equine Geneticist will depend on college enrollment.

Related Occupations

Occupations related to the Equine Geneticist include: Equine Nutritionist, Laboratory Animal Technician, and Veterinarian.

EQUINE JOURNALIST

Description

Equine Journalists utilize the written word to communicate. They create nonfiction works for horse-related books, magazines, trade journals, newspapers, newsletters, radio and television programs, advertisements, and motion pictures. Most Journalists do not write fiction, as this is a specialized area. Equine Journalists are required to gather information through personal interviews and research, to select and organize the horse-related material, and to put it into words that are easily understood by the reader. They should be able to make scientific and technical information concerning the horse easy to understand by the average person. Equine Journalists may be employed as staff writers for specific horse publications or as independent freelance writers. As independent freelance writers they submit their work to various publications for a fee.

Educational and Training Requirements

A high school education is required, along with a four-year degree in Journalism, Language Arts, or Communications and courses in Liberal Arts. Practical experience as well as a broad background in the horse industry are essential for this profession. Practical writing experience with high school and college newspapers as well as community newspapers is also beneficial.

Personal Qualifications

The Equine Journalist should have a genuine passion for writing and be able to express ideas logically and clearly. Other beneficial traits for the Equine Journalist include creativity, self-motivation, and perseverance. Irregular hours are considered the norm for this occupation. Most work is performed indoors in an office setting, but travel is also an essential part of the Equine Journalist's routine. Equine Journalists must have excellent communication and computer skills in order to succeed as professional writers. Finally, they must be able to work independently and to handle pressure when meeting deadlines.

Licenses and Certification

There is no formal licensing or certification for the profession of Equine Journalist.

Employment Outlook

The outlook for employment as an Equine Journalist is expected to be highly competitive because a large number of people are attracted to this field. The increase in the number of horse-related publications in recent years has resulted in an increased demand for professional Equine Journalists. The need for technical Equine Journalists is expected to increase in the future due to the increased expansion of scientific and technical information concerning the equine athlete.

Related Occupations

Occupations related to the Equine Journalist include: Equine Advertising Specialist, Equine Artist, Equine Book Dealer, and Horse Industry Secretary.

EQUINE LIBRARIAN

Description

The function of the Librarian specializing in the subject of horses is to assist library patrons in locating information about horses. The Librarians oversee the selection and organization of materials pertaining to the horse as well as a variety of other materials normally found in a library. They are responsible for managing the library staff as well as the overall operation of the library, which includes services to the public, technical services, and administrative services. Librarians may be responsible for the acquisition of books, periodicals, and audiovisual materials pertaining to the horse and other subjects. They prepare budgets, supervise programs sponsored by the library, and handle all public relation matters. Librarians are employed by public libraries, government libraries, school and university libraries, museum libraries, and research libraries.

Equine Librarians are usually employed by large libraries dedicated to the horse or a specific horse breed association. In addition to their usual duties as Librarians, they may be responsible for arranging and providing meeting rooms for various horse organizations, assisting teachers in obtaining instructional material, preparing bulletin boards and exhibits within the library, directing library patrons to standard references, organizing and maintaining equine periodicals, preparing invoices, cataloging and coding all library materials, and instructing patrons on the use of the computer database. School libraries employ Librarians to teach students to use the library and media center. Other Librarians are employed by the government, professional horse breed associations, equine museums, and veterinary hospitals and clinics. Research laboratories employ Librarians where they are required to prepare articles, conduct literary searches, and compile biographies. Positions may be full- or part-time. Work may be performed during the evenings and weekends in addition to the normal 40-hour week.

Educational and Training Requirements

A master's degree in Library Science (MLS) is required for the professional Librarian. The American Library Association has accredited at least 50 colleges in the United States with an MLS program. Entry-level positions at a library may require an applicant to have an associate degree in Library Technology. Courses in Liberal Studies as well as library-related study are also required for this position. College courses should include library science, history of books, censorship, printing, computer science, cataloging, classification, library administration, and automation technologies. Any practical experience or volunteer work within a library is beneficial for this profession, and a general knowledge of horses is helpful but is not a requirement.

Personal Qualifications

A genuine interest in horses is essential for the Librarian specializing in equine studies. Practical experience with horses is not a requirement but is beneficial in meeting the needs of the library patron. All work is performed indoors in an office setting. Most of the Librarian's time is spent working at a desk in front of a computer terminal, so the Librarian must be computer literate. Equine Librarians are required to have various communication skills in order to deal with the general public; in addition, they must have a pleasant personality and a great deal of patience when performing their daily duties.

Licenses and Certification

Some states require public school Librarians to be certified as teachers with courses in Library Science before becoming certified Librarians. Some states require certification of public Librarians employed by a county or state library system.

Employment Outlook

The employment outlook for the position of Librarian is limited because of the trend toward reducing the overall budgets of public libraries. School libraries will continue to eliminate staff in order to reduce expenses. In addition, the increase in the use of computers in all library operations will reduce the demand for the professional Librarian.

Related Occupations

Occupations related to the Equine Librarian include: Equine Educator and Pedigree Researcher.

EQUINE LOAN OFFICER

Description

The duty of the Loan Officer specializing in the horse industry is to examine and underwrite applications for loans or lines of credit. Loans are a major source of income for financial institutions. People in the horse industry may take out a *mortgage loan* to purchase a breeding farm or training center, or a *commercial loan* to buy equipment such as tractors, trucks, and horse trailers. Finally, there are some financial institu-

tions that offer a *consumer loan* for the purpose of purchasing horses.

Loan Officers interview applicants to determine their ability to pay debts. They check applicants' references, credit ratings, sense of business, collateral, capital, and character, and use the information they obtain to determine whether an applicant will be refused or accepted as a borrower. If the loan is granted the terms, the interest rates and the repayment schedule are set by the Loan Officer. In some cases, Loan Officers are involved in foreclosure proceedings.

Educational and Training Requirements

A high school education and a two- or four-year college degree in Accounting, Business Administration, or Finance are essential for this position. College courses should include economics, business law, and computer science. A general knowledge of the horse industry is beneficial but is not a requirement.

Personal Qualifications

Equine Loan Officers should possess good communication skills in order to deal with clients, as well as a good sense of judgment. They should also be understanding, discreet, tactful, patient, and compassionate; and they should be able to work independently with little or no supervision. Most of the work is performed indoors in an office setting, but sometimes the Loan Officer must visit a property site or attend a public horse auction. Most Loan Officers work 8 to 10 hours per day. Loan Officers working within the horse industry must have insight into the various types of horse-related businesses that make up the industry; therefore, they should attend equine seminars and conferences to develop and increase their knowledge of the horse industry.

Licenses and Certification

The American Institute of Banking offers courses through correspondence, and through some college and universities, to Loan Officers and students interested in a career in banking. Completion of these courses leads to the title of Certified Lender in Business Banking. This certification enhances the possibility of employment as a Loan Officer.

Employment Outlook

As commerce, investment, and trade within the horse industry become international in scope, the Loan Officer will be in demand. The expansion of both domestic and international services, stricter laws on loans and taxes, and the increased use of computers will also increase the need for the experienced Loan Officer.

Related Occupations

Occupations related to the Equine Loan Officer include: Equine Accountant or Bookkeeper, Equine Real Estate Broker and Agent, Horse Insurance Agent or Broker, and Horse Insurance Underwriter.

EQUINE MASSAGE THERAPIST

Description

The Equine Massage Therapist massages the body of the horse by using techniques such as kneading, rubbing, and stroking the flesh and muscle tissue. Massage therapy is designed to allow the horse to be as athletic as possible. Various types of riding disciplines require the various types of muscle groups to be supple and flexible. Massaging aids in stimulating blood circulation, relaxing contracted muscles, and facilitating the removal of waste products from the body, and it helps to relieve other physical conditions as well. By utilizing their hands or vibrating equipment, Equine Massage Therapists develop and restore functions, prevent loss of physical capacities, and maintain the optimum performance of the horse. They may also give horse owners directions in remedial exercises for horses or examine horses and recommend body conditioning activities and treatments. (See Figure 10.5.)

FIGURE 10.5
The Equine Massage Therapist allows the horse to be as athletic as possible.

Educational and Training Requirements

A high school education and vocational training in Equine Science are essential, as well as professional training in Massage Therapy from a certified equine massage program with courses in equine anatomy, physiology, and behavior. Any training in basic horsemanship is also very helpful.

Personal Qualifications

Equine Massage Therapists should possess strong interpersonal skills so they can help their clients understand the recommended treatments. They should also be compassionate and possess a strong desire to help the equine patient. Equine Massage Therapists should have a high degree of manual dexterity and physical strength in order to treat their equine patients. They must be able to stoop, kneel, crouch, lift, and stand for long periods of time. They must also be willing to travel long distances to service their clients. Their work is usually performed indoors inside a barn or stable.

Licenses and Certification

There is no mandatory licensing requirement for the Equine Massage Therapist, although those who treat racehorses on the racetrack must be licensed by the state racing commission in order to gain access to the stable area on the racetrack grounds. Certification may be acquired through various massage therapy training organizations such as the Equissage Myotherapist Association.

Employment Outlook

The employment outlook for the Equine Massage Therapist is good. This is a new and growing field, and there is a growing acceptance within the horse industry of this practice and its application to equine sports training. As massage therapy becomes increasingly popular, there will be an increased demand for trained Equine Massage Therapists.

Related Occupations

Occupations related to the Equine Massage Therapist include: Equine Chiropractor, Equine Veterinary Assistant, Groom, and Horse Trainer.

EQUINE NUTRITIONIST

Description

Equine Nutritionists develop, test, and promote various types of feed rations for the horse. They scientifically evaluate prepared horse feeds based on their texture, appearance, flavor, and nutritional value, and make modifications in the diet of the horse. They are responsible for recording the various amounts and types of ingredients as well as test results. Equine Nutritionists also develop new products and participate in product improvement and promotions for feed companies. They attend equine nutrition seminars and conventions in order to accumulate the latest research data, and they confer with veterinarians and other professionals to coordinate the medical and nutritional needs of the horse.

Equine Nutritionists provide services to horse-breeding operations, racing stables, and individual horse owners. They perform nutritional screening for their clients and offer advice. They assess the nutritional needs of their clients' horses, develop and implement nutritional programs, and evaluate and report the results. Some are employed by large animal feed companies or by universities as part of animal nutrition research teams.

Educational and Training Requirements

The basic educational requirement is a bachelor of science degree in Animal Science or Equine Science. However, most Equine Nutritionists possess a doctorate degree in Animal Nutrition with an emphasis on the equine species. College courses should include animal nutrition, biology, microbiology, and physiology. Other beneficial courses are mathematics, business, statistics, computer science, psychology, sociology, and economics.

Personal Qualifications

The Equine Nutritionist should have intelligence, an interest in science, and the patience to work with details. Most Equine Nutritionists have regular hours while working under laboratory conditions but work long hours in the field when conducting special studies, performing special tests, or completing projects. Equine Nutritionists must be willing to travel long distances in order to meet with clients. They must have good communication skills, both written and oral, when dealing with horse owners, feed manufacturers, veterinarians, and research directors. They also must be able to work independently and handle responsibility.

Licenses and Certification

There is no mandatory licensing or certification requirement for the Equine Nutritionist.

Employment Outlook

The employment outlook for the Equine Nutritionist is favorable. Modern horse owners are becoming more knowledgeable in matters concerning the overall health of their horses. Most horse people are well informed about the role nutrition plays in the physical development, athletic ability, and daily maintenance of their horses. This nutritional awareness on the part of horse owners creates a definite need for the professional Equine Nutritionist. Also, as feed companies expand product lines, the need for Equine Nutritionists will become greater.

Related Occupations

Occupations related to the Equine Nutritionist include: Feed and Grain Dealer, Laboratory Animal Technician, and Veterinarian.

EQUINE PHOTOGRAPHER

Description

The Equine Photographer uses the camera to photograph the equine subject in order to sell a product, capture a special moment, provide entertainment, or recall a memory. Creating commercial quality photographs requires technical expertise and creativity. There are some professional photographers who develop and print their own photographs; however, this practice requires a fully equipped dark room and the technical skill to operate it. Most Equine Photographers specialize in commercial, portrait, or media photography in a particular segment within the horse industry. You will find photo specialists in racing, hunting, jumping, and dressage. Photography is considered to be an art medium; therefore, some photographers sell their photographs as artwork, placing an emphasis on creativity and self-expression.

Educational and Training Requirements

A high school education is required plus a two- or four-year degree in Photography with courses in Equine Science emphasizing conformation, anatomy, and animal behavior. Many photographers enhance their expertise by attending horse-related seminars. College courses in business skills are also beneficial for this profession. Knowledge of mathematics, physics, and chemistry is essential in understanding the functions of film, lens, lighting, and the photo development process.

Personal Qualifications

Equine Photographers must have a genuine love of the horse. They must possess artistic and imaginative qualities as well as excellent eyesight and manual dexterity. Other important traits of the Equine Photographer include patience, accuracy, and an orientation to detail. Equine Photographers should be able to work independently as well as with others, and they must have good communication skills when dealing with both horses and clients. Finally, they must be willing to travel and carry extensive photo equipment wherever they go.

Licenses and Certification

There is no formal licensing or certification for the profession of Equine Photographer.

Employment Outlook

As more and more people purchase and show horses, the demand for the professional Equine Photographer should increase in the near future. The growing demand for visual images in the horse industry in areas such as education, entertainment, marketing, and communication should spur the demand for the Equine Photographer. Equine photography, however, is a highly competitive field, and only those Equine Photographers who are the most skilled and have developed the best reputations in the horse industry are able to develop enough work to support themselves.

Related Occupations

Occupations related to the Equine Photographer include: Equine Advertising Specialist, Equine Artist, and Equine Journalist.

EQUINE REAL ESTATE BROKER AND AGENT

Description

The duty of the Equine Real Estate Broker and Agent specializing in the horse industry is to bring landowners and buyers together to negotiate the sale of real estate. Such real estate sales include the sale of land utilized for horse farms and stables. Equine Real Estate Brokers and Agents locate buyers and renters for farms, vacant land, and established horse facilities. For these services, they receive a percentage of the selling price of the property. A Real Estate Broker is an independent Realtor licensed to sell, rent, and appraise properties. The Real Estate Agent is a salesperson who offers services to a licensed broker through a contractual agreement. Agents must be licensed in the state in which they are conducting business. They receive a portion of the commission earned in the transaction. Equine Real Estate Brokers and Agents take prospective buyers to physically view properties that meet their equestrian needs and income and negotiate the terms of the sale to meet the needs of both parties. (See Figure 10.6.)

FIGURE 10.6
Real Estate Brokers and Agents bring landowners and horse owners together to negotiate the sale of real estate. (Courtesy of PhotoDisc)

Educational and Training Requirements

For this profession, a high school education is required, along with specialized training in the field of real estate. A college degree is beneficial but is not a requirement. College courses should include real estate, economics, financing, computer science, and business management. Colleges and universities now offer real estate programs leading to a degree or a certificate in the field of real estate. There are some state universities that offer home-study real estate programs.

Personal Qualifications

Equine Real Estate Brokers and Agents must have good communication skills when dealing with people on a daily basis. They must be able to work independently with little supervision. Most work is performed indoors in an office setting, although when showing property to prospective clients, they may be required to walk about the property to determine its boundaries, water and electric lines, and overall terrain. They must be willing to work long hours including evenings; a 7-day workweek is not uncommon in this profession. Finally, they must be willing to drive long distances in order to accommodate their clients.

Licenses and Certification

All states require a formal license for Real Estate Brokers and Agents. An applicant must be at least 18 years of age and a high school graduate. Licensing requires the applicant to pass a written examination. The written examination usually covers basic real estate transactions and all laws pertaining to the sale of property. Some states require Real Estate Agents to accumulate 30 hours of classroom instruction in real estate before a license is issued. A Real Estate Broker must complete 90 hours of formal classes and one to three years of selling real estate before a license can be issued.

Some states require all Real Estate Agents to obtain a bond guaranteeing their performance of a contract or obligation. In some states, the Real

Estate Broker or Agent must compile 60 hours of college credit before a license is issued. Once Real Estate Brokers and Agents are licensed, they are required to take continuing education courses in real estate in order for their license to be renewed.

Employment Outlook

The employment outlook for the Equine Real Estate Broker and Agent is good for the near future as horse ownership increases. The increase in horse ownership will result in an increase in the demand for rural property to provide for these horses. Employment, however, also depends on the economy of the horse industry; when the economy is in the down trend, horse property sales will also decline.

Related Occupations

Occupations related to the Equine Real Estate Broker and Agent include: Appraiser, Equine Accountant or Bookkeeper, Equine Attorney, and Horse Insurance Agent or Broker.

EQUINE RECREATION DIRECTOR

Description

Equine Recreation Directors organize and administer all horse-related activities affiliated with summer camps, resorts, amusement parks, and hotels. They assist individuals and groups in riding lessons, trail rides, grooming lessons, horse shows, rodeos, and hay rides. They are responsible for planning, promoting and organizing all horse-related recreational activities. Equine Recreation Directors are also responsible for the hiring and training of both salaried staff members and volunteers, for the daily care and maintenance of the horses utilized in the recreational programs, and for public relations, budgets, bookkeeping, and paying bills that are incurred.

Educational and Training Requirements

A high school education is essential, as well as a college degree in Equine Science, Parks and Recreation Management, or Physical Education, with courses in Recreation and Leisure Studies. Students should acquire a working knowledge of horses, equitation, and horsemanship, as well as athletics. Work experience should include several years as a recreational volunteer.

Personal Qualifications

The full-time professional Equine Recreation Director should have a genuine love for horses and people. Good communication skills are required, especially for dealing with employers, employees, and the general public. Equine Recreation Directors must also have the ability to teach, plan, take charge, and cooperate. They must be able to react quickly in the event of an emergency situation, and they must have an even temper, patience, and self-confidence.

Licenses and Certification

There is no formal licensing requirement for Equine Recreation Directors; however, they may become certified through various organizations such as the American Riding Instructors Program, the Association for Horsemanship Safety and Education, and the Horsemanship Safety Association.

Employment Outlook

This profession is expected to grow in the coming years, as the general population is experiencing more leisure time as well as more income, which will create a demand for more horse-related recreational activities. Those with a graduate college degree will have the best opportunities for obtaining available positions.

Related Occupations

Occupations related to the Equine Recreation Director include: Equine Educator, Mounted Park Ranger, and Riding Instructor.

EQUINE SPORTSCASTER

Description

Equine Radio and Television Sportscasters present news, sports, commercials, interviews, and reports on horse-related activities of interest to their audiences. They may be involved in creating written scripts for their presentations. They are responsible for presenting horse-related news stories, and for introducing videotaped topics and live transmissions from specific locations such as racetracks, horse shows, rodeos, and polo events. The Equine Sportscaster must conduct live interviews with horse owners, trainers, jockeys, and other professional horse people.

Educational and Training Requirements

A high school education is mandatory for this position. While in high school, courses should include public speaking, language arts, and drama. A degree in Broadcast Journalism from a college or a technical broadcasting school is desirable. College courses should include a foreign language, computer science, and electronics. Students may obtain valuable training at a college campus radio or television studio. Many cable systems and radio stations offer on-the-job training programs for students. Any knowledge of horses or horsemanship is beneficial to the Sportscaster specializing in equine events.

Personal Qualifications

The Equine Radio and Television Sportscaster must have a good speaking voice with excellent pronunciation and correct use of the English language. For television work, the Equine Sportscaster must have a neat, appealing appearance as well as a good voice. Equine Sportscasters must be computer literate, as scripts are written and edited by utilizing a computer. They work both indoors, in soundproofed studios, and outdoors when covering specific horse-related events. When working outdoors, they are exposed to all types of weather conditions. Equine Sportscasters commonly work long, irregular hours and under the stress and pressure of a tight schedule.

Licenses and Certification

Any Sportscaster or Announcer who is required to operate broadcast transmitters in radio and television stations must obtain a Federal Communication Commissions (FCC) restricted radio telephone operator permit.

Employment Outlook

The employment outlook for the Equine Sportscaster is limited in scope, as this is a small, highly competitive field. The position usually attracts more job seekers than there are jobs available. Specialization is found primarily in larger stations and networks, although many minor stations now seek specialists as well. Because such a small number of equine events attract large radio and television audiences, the need for the professional Equine Sportscaster is limited.

Related Occupations

Occupations related to the Equine Sportscaster include: Announcer, Auctioneer, and Equine Advertising Specialist.

EQUINE TRANSPORTATION SPECIALIST

Description

The Equine Transportation Specialist is responsible for arranging, coordinating, and directing the transportation of horses throughout the United States and the entire world. The transportation of horses, both domestically and internationally, is subject to special laws and regulations. To be sure that shipments of horses move smoothly, safely, and legally, many shipping companies specialize in the transportation of horses; these companies transport horses by either land, sea, or air. Equine

Transportation Specialists know the best method to ship horses and all the aspects of the ports to which horses must travel. They work with commercial freight firms as well as trucking and airline firms. They are responsible for the paperwork involving customs, shipping requirements, licenses, and insurance. They may work for large international equine transporting firms or act as independent brokers, handling either the import or export of horses for clients in the United States and abroad. An Equine Transportation Specialist is involved in working with a variety of professionals such as shipping dispatchers, export and import agents, customs agents, import managers, and truck and van drivers.

Educational and Training Requirements

A high school education is required with courses in business, economics, computer science, and marketing. Fluency in a second language is particularly beneficial; the ability to communicate in German, French, Spanish, and Arabic is a definite asset for the Equine Transportation Specialist. A college degree in Business Management with courses in accounting, marketing, foreign trade, political science, world geography, finances, and international economics is applicable for this profession.

Personal Qualifications

Equine Transportation Specialists should be outgoing, intelligent, and enjoy working with people. They must have an analytical mind in order to handle all the details involved in transporting horses. They should be able to work independently with little supervision, and they must be willing to travel a great deal on short notice. They must also have good communication skills, especially when dealing with clients, custom agents, airline officials, and import and export agents. A basic knowledge of horses and horsemanship is recommended for this profession but is not a requirement.

Licenses and Certification

The United States Department of Commerce offers licenses to firms specializing in exports. The tests for this type of license include subjects such as laws, procedures, and documentation. For details concerning licensing, contact the nearest branch office of the United States Department of Commerce.

Employment Outlook

The employment outlook for the Equine Transportation Specialist is good. Today many horses are shipped all over the world for showing, breeding, and racing. As more and more horses are competing on an international level, the need for the professional Equine Transportation Specialist will increase.

Related Occupations

Occupations related to the Equine Transportation Specialist include: Bloodstock Agent, Equine Attorney, and Horse Insurance Agent or Broker.

EQUINE TRAVEL AGENT

Description

Travel Agents specializing horse-related trips give advice to their clients who are planning trips and arrange for transportation, hotel accommodations, automobile rentals, tours, and equestrian activities. They also advise clients on weather conditions, restaurants, and tourist sites. For foreign travel, they become involved in customs regulations, required documents, passports, visas, birth certificates, vaccination certificates, and currency exchange rates. Equine Travel Agents sometimes make presentations to horse organizations and horse-related groups in order to promote their services.

Educational and Training Requirements

The minimum requirement for the Equine Travel Agent is a high school education and attendance at a vocational trade school, either public or private, offering a Travel Agent program. Only a few colleges and universities offer a bachelor's or master's degree in Tourism and Travel. High school and college courses should include geography, computer science, foreign languages, world history, hotel reservations, automobile reservations, general insurance, and communications. Good typing, writing, and oral communications skills are desirable qualifications for this position. A correspondence course in Tourism and Travel is available through the American Society of Travel Agents (ASTA). A basic knowledge of horses and horsemanship is beneficial but is not a requirement.

Personal Qualifications

The Equine Travel Agent must be willing to work long, irregular hours that include weekends and holidays. Most work is performed indoors within a general office setting. It is here that the Equine Travel Agent deals with clients, paperwork, airlines, hotels, and automobile rental agencies in making travel arrangements and promoting group tours for horse people. Equine Travel Agents must be well traveled so they are able to relay to their clients their personal knowledge about a particular city or country, and they must possess good marketing skills as well as patience when dealing with clients. Finally, they must enjoy working with people.

Licenses and Certification

There is no federal licensing requirement for Travel Agents. The following nine states require registration or certification of all Travel Agents regardless of their area of specialization: California, Florida, Hawaii, Illinois, Iowa, Ohio, Oregon, Rhode Island, and Washington. The Institute of Certified Travel Agents (ICTA) offers courses to experienced Travel Agents that lead to the designation of Certified Travel Counselor (CTC). New programs sponsored by the ICTA include courses to become a Certified Travel

Agent (CTA) and the Travel Agent Proficiency (TAP) Test.

Employment Outlook

As household incomes increase, the amount of money each person or family spends on travel is expected to increase in the near future. The number of elderly people who travel is growing, as well as the number of people who travel on vacations. This will result in an increased demand for the professional Equine Travel Agent. Also, there is a growing interest and demand on the part of horse people for special equestrian-oriented tours and vacations to foreign countries and in the western part of the United States. It is the role of the Equine Travel Agent to arrange these horse-related vacations and tours. However, the growing use of personal computers that allow people to arrange their own travel plans may tend to reduce the employment opportunities for all Travel Agents in the future.

Related Occupations

Occupations related to the Equine Travel Agent include: Equine Accountant or Bookkeeper, Equine Real Estate Broker and Agent, and Horse Insurance Agent or Broker.

EQUINE VETERINARY ASSISTANT

Description

The primary duty of the Equine Veterinary Assistant is to prepare horses for surgery, provide postoperative care, and administer medication to equine patients. Work is usually performed under the direct supervision of a licensed veterinarian. Equine Veterinary Assistants also prepare the treatment room for the examination of horses and hold or restrain horses during examination and treatment. In addition, they administer injections, apply wound dressings, take and record vital signs (temperature, pulse, and respiration), assist veterinary surgeons during surgical procedures, assist in taking x-rays, and communicate with the owners of horses. In general, Equine Veterinary Assistants may be assigned direct responsibility for any activity in a veterinary practice except diagnosis, treatment, and surgery.

Educational and Training Requirements

A high school education and vocational training in Equine Science are essential for this profession. A two- or four-year degree in Animal Science, Veterinary Technology, or a related field is also beneficial. Finally, any knowledge of general horsemanship skills is helpful.

Personal Qualifications

Some of the personal traits required for Equine Veterinary Assistants are patience, compassion, and the ability to be a team player. They must enjoy working with both horses and people, and they must have a pleasant personality and good communication skills when dealing with clients, employers, and fellow employees. Equine Veterinary Assistants must be willing to work long hours each day, as well as a 6-day week, and must be willing to put in extra hours if needed and to be on call in the event of an emergency situation. Most duties are performed indoors in a veterinary hospital or clinic, although sometimes they may be required to work outdoors at a client's stable or farm.

Licenses and Certification

Most states in the United States require Veterinary Assistants to be registered or certified. Applicants should follow the guidelines for certification that are applicable in their state of residence or employment.

Employment Outlook

Veterinary medicine is considered to be a career with great potential. Horse ownership is increasing, as many people are moving out of the cities and into the suburbs. Rural veterinary practitioners will also thrive because horse owners are exhibiting greater scientific interest in the breeding, training, and exhibiting of horses. Naturally, this will result in an increased need for equine veterinarians as well as Equine Veterinary Assistants.

Related Occupations

Occupations related to the Equine Veterinary Assistant include: Equine Dentist, Laboratory Animal Technician, and Veterinarian.

EQUINE VIDEOGRAPHER

Description

The duty of the Equine Videographer is to videotape horse-related events such as horse shows, races, rodeos, and polo matches. Equine Videographers utilize video cameras that provide adjustment settings for both close and distance recording of these events. In addition to the video camera, other tools may include editing equipment, tripods, lighting fixtures, microphones, filters, and lens. The Equine Videographer may also produce instructional videos pertaining to the horse industry. Some of the horsemanship skills available to audiences on video include riding, training, grooming, health, conformation, nutrition, and many other equine topics as well. Equine Videographers are also used to provide videos for the sale of horses, stallion promotions, and real estate sales. This profession requires long hours of shooting, editing, and duplicating tapes with short deadlines.

Educational and Training Requirements

A high school education or vocational training in video production is essential. Equine Videographers may also obtain training on a college level by attending a university, community college, or a private trade school offering courses in camera work, editing, and video production. Basic college courses in Videography focus on equipment, processes, techniques, editing, and business. Business courses should include bookkeeping and computer science, which are essential for invoicing and editing. A basic knowledge of horses and horsemanship is a definite requirement for this position.

Personal Qualifications

Equine Videographers should possess certain characteristics, such as imagination, creativity, patience, and an orientation to detail. They must have good eyesight, manual dexterity, and an artistic mind. They should enjoy working with both people and horses. They must be able to work independently with little or no supervision, and they must have good communication skills in order to deal with employers, clients, and fellow employees. Their work is performed indoors in studios as well as outdoors on location. The Equine Videographer must often work long, irregular hours and must be available to travel to various locations in order to meet with clients and subjects. A prospective Equine Videographer should put together at least two sample reels of work to show to prospective clients.

Licenses and Certification

There is no formal licensing or certification requirement for the profession of Equine Videographer.

Employment Outlook

The employment outlook for the Equine Videographer is poor because there are more individuals who wish to become Equine Videographers than there is employment to support them. Competition is very keen in this new and exciting field. Those who have the most skills, the best business ability, and the best reputations in the field are able to find salaried positions or successfully start their own businesses. Many Videographers have full-time jobs in other fields and take videos of horse-related events on weekends. Equipment is very expensive. There are enormous start-up costs because of the expense of purchasing new digital cameras and the high editing facility costs. It should be noted that the field of Videography is part of the communications industry, which is the largest industry worldwide.

Related Occupations

Occupations related to the equine Videographer include: Equine Artist, Equine Journalist, and Equine Photographer.

EXERCISE RIDER

Description

Exercise Riders are responsible for riding the racehorse during the morning exercise sessions on the racetrack. They monitor the horse's movement, temperament, and physical condition. They report to the trainer on the progress or decline in the horse's condition on a daily basis and may make suggestions to the trainer concerning changes in equipment or the training schedule. The Exercise Rider also works with the trainer in getting a racehorse accustomed to the starting gate. Some Exercise Riders are employed on a farm or training center and specialize in the development of the young horse preparing for a racing career. Exercise Riders are also responsible for the cleaning and conditioning of the track after each use. They are asked to assist in daily stable chores such as walking, washing, feeding, and grazing horses. An Exercise Rider may be employed by one trainer for a set weekly salary or as a freelance Exercise Rider working for several trainers on a set fee for each horse ridden. An Exercise Rider employed by one trainer may ride three to six horses per day, while freelance Exercise Rider may ride as many as eight to ten horses each day. (See Figure 10.7.)

Educational and Training Requirements

The basic educational requirement for the Exercise Rider consists of a high school diploma and practical riding experience. A basic knowledge of horsemanship skills is essential for this profession. A college education in the field of Equine Science is not a requirement but may be beneficial. A knowledge of horse behavior is also helpful when dealing with young inexperienced horses.

FIGURE 10.7
An Exercise Rider keeps the racehorse physically fit with daily exercise. (Photo courtesy of Michael Dzaman)

Personal Qualifications

The Exercise Rider must have a genuine love for the horse. Basic riding skills are essential, as well as personal characteristics such as being gentle, firm, calm, confident, and most of all patient. Good communication skills are important when dealing with employers and fellow workers. Exercise Riders must be physically fit and must not weigh more than 135 pounds. They must be able to work 6 days per week in the early hours of the morning in all types of weather conditions. Finally, they must be willing to travel to various racetracks throughout the country.

Licenses and Certification

All Exercise Riders must obtain a license at the racetrack where they are employed. The licensing procedure entails fingerprinting, photographs, and payment of the required fee to the state racing commission. Exercise Riders employed at training centers and farms are not required to be licensed.

Employment Outlook

The overall decline in horse racing in America will result in limited opportunities for the Exercise Rider. However, employee turnover runs high in this type of work, which results in a constant demand for the professional Exercise Rider. Racing schedules also affect employment, as some racetracks do not operate year-round.

Related Occupations

Occupations related to the Exercise Rider include: Groom, Horse Trainer, Jockey, and Riding Instructor.

FARM EQUIPMENT DEALER

Description

Many large horse farms require specialized equipment and machines for their daily operations. Horse farms have several types of tractors with various levels of horsepower. In addition to tractors, horse farms may utilize hay balers, treadmills, starting gates, manure spreaders, mowers, walking machines, sprayers, and irrigation equipment. The duty of the Farm Equipment Dealer is to sell and service farm equipment and machinery to the horse farm or ranch owner. Some Farm Equipment Dealers represent only one manufacturing company and sell and service only that company's products. Others are independent dealers who will sell and service various farm equipment manufacturers' products to their customers. Most dealers are located in rural farm areas. They usually display their new and used equipment on a lot located within city limits or just outside of a village or town. They deal directly with customers, manufacturers, mechanics, and service technicians. (See Figure 10.8.)

FIGURE 10.8
The Farm Equipment Dealer sells and services farm equipment for the horse owner.

Educational and Training Requirements

The educational requirements for this position consist of a high school education plus a two- or four-year degree in Agriculture, Agronomy, Agribusiness, Agricultural Mechanics, or Animal Science. College courses should include accounting, marketing, computer science, and economics. Practical experience at a farm equipment dealership as a mechanic or general farm experience is desirable. Most Farm Equipment Dealers learn their trade by completing an apprenticeship program offered through farm equipment manufacturers, which may last from two to three years.

Personal Qualifications

Farm Equipment Dealers must have good communication skills, both oral and written, in order to deal with customers, mechanics, technicians, and farm equipment company representatives. They must be willing to work long hours, including weekends and holidays. As with other business persons, they must be dependable, fair, trustworthy, and reliable. Most of their work is performed indoors, but on occasion they must meet with clients in an outdoor setting in all types of weather conditions.

Licenses and Certification

There is no licensing or certification requirement for the position of a Farm Equipment Dealer. In some cases, the dealer becomes an authorized dealer for a particular company.

Employment Outlook

With farmlands being consolidated into fewer but larger farms and with the increasing use of new farming procedures, most horse owners and ranchers are investing in modern, more efficient, and specialized equipment. The Farm Equipment Dealer provides this equipment.

Related Occupations

Occupations related to the professional Farm Equipment Dealer include: Feed and Grain Dealer, Groom, Horse Trailer and Van Dealer, and Retail Tack Shop Operator.

FARM OR RANCH MANAGER

Description

The Farm or Ranch Manager may manage a herd of horses, including broodmares, stallions, yearlings, weanlings, and foals. Duties may include production of field crops, such as hay and grain, grown on the property as part of the horse operation. Duties may also include breeding, feeding, maintaining the health of the herd, transporting, and marketing of the horses. The major responsibilities of the Farm Manager are supervising the work of all employees, maintaining all records pertaining to the operation, and maintaining all equipment and buildings on the farm or ranch. The daily work is usually performed outdoors, and the working hours are usually irregular, although hours vary depending on the type of horse operation (breeding, boarding, or training) and the season of the year.

Educational and Training Requirements

A high school education plus a vocational program emphasizing Agriculture and Livestock Production would be desirable. A thorough knowledge of agriculture and business management is beneficial. An associate degree or a bachelor of science degree in Agriculture, Animal Science, Business Management, Equine Science, or Computer Skills would also be an advantage to the prospective Farm Manager.

Personal Qualifications

A prospective Farm Manager must be able to work closely with people. Farm Managers must be able to make managerial decisions related to the daily operations of a horse farm, and they must have a broad scope of horse experience, a pleasing personality, and good communication skills. The Farm Manager is usually required to work long hours as well as weekends.

Licenses and Certification

There is no licensing or certification required for the position of Farm or Ranch Manager. However, a specific driver's license may be required for operating various vehicles used to transport horses.

Employment Outlook

Recent optimistic trends indicate an increase in the popularity and demand for light horses, particularly for recreation. If these trends continue, horse production will rise with demand, thus resulting in an increase in the number of horse farms and managerial positions.

Related Occupations

Occupations related to the Farm or Ranch Manager include: Groundskeeper and Horse Trainer.

FARRIER

Description

There are two basic types of Farriers. The common Farrier selects the types of shoes and fits,

shapes, and nails or glues the shoes on the horse's hooves. The second type of Farrier is the Corrective Shoer. This Farrier specializes in providing specially designed shoes to correct a particular problem plaguing a horse. These problems can include a faulty gait, serious ailments, or recovery from surgery. Both types of Farriers examine the hooves for any deformities or ailments, and they trim and shape the foot to its proper balance and level. They may perform cold shoeing whereby aluminum shoes are shaped to fit the hooves without heating the shoes, or they may perform hot shoeing whereby a steel or steel alloy shoe is heated in a furnace in order to shape the shoe to properly fit the hoof. (See Figure 10.9.)

Educational and Training Requirements

A Farrier should have a high school education with vocational training in Farrier Science or a two- or four-year degree in Equine Science. Courses in equine behavior, anatomy, and business management are beneficial. A good background in equitation skills as well as good horsemanship skills are essential for this position. Finally, an apprenticeship period with a working Farrier is essential before entering this field. The knowledge and skill so necessary for

FIGURE 10.9
The Farrier is responsible for health of the hooves as well as providing shoes. (Courtesy of Kentucky Horse Park)

this profession may be self-taught or acquired over a long period of time by practice, observation, and serving as a trainee to a skilled Farrier.

Personal Qualifications

Farriers must have a genuine passion for this type of work with horses. They must be willing to work long hours each day, and a 5- or 6-day week. They must be willing to drive on a daily basis to various worksites. Farriers must be self-starters with a high degree of business sense, and they must have good communication skills with both horses and clients. Most of their work is done outdoors in all types of weather conditions. Farriers must be in excellent health and physically fit in order to perform their work and handle unruly horses. Farriers must have good eyesight, hand-eye coordination, hand and finger dexterity, agility, and a good sense of balance and of the angles related to the conformation. An instinctiveness about approaching horses and inventiveness are desirable attributes of an aspiring Farrier. The Farrier must be willing and able to work hard, under trying conditions, for long periods of time.

Licenses and Certification

There are no existing laws requiring licensing or certification to be a Farrier; however, the American Farrier's Association (AFA) has a program that certifies Farriers at various skill levels. The two levels of AFA certification are the AFA Certified Farrier and the AFA Certified Journeyman Farrier. An Intern classification is also available as an initial step toward certification. Certification study and testing leads to improved knowledge and skill levels for the Farrier.

Intern Classification. Farriers who have just completed Farrier school may opt to try for Intern classification status. A written test and a practical test ensure that the Farrier can shoe a horse within a specified time period and to a published standard.

AFA Certified Farrier. Farriers who have been shoeing for a minimum of one year may opt to take the test for AFA Certified Farrier. In addition to the written and practical tests, the AFA Certified Farrier completes a set of sample shoes with various modifications, which show the ability to do the basic work required of Farriers.

AFA Certified Journeyman Farrier. Farriers who have been shoeing for a minimum of two years may take the test for the AFA Certified Journeyman Farrier. There is a more comprehensive test on anatomy, gaits, and shoeing and a set of sample shoes. A practical test requires the Farrier to shoe a horse with handmade shoes to a set standard in 2 hours.

Employment Outlook

Given the present increase in horse ownership throughout the United States, there will be an increased demand for the professional Farrier. Competition is high in some parts of the country due to the concentration of horses in these areas. In other parts of the country where the horse population is less concentrated, a Farrier must travel greater distances in order to service clients. A dedicated, competent Farrier is in demand almost everywhere in the United States where there is a horse population. Since most horses should be trimmed or shod every 6 to 8 weeks year-round, there is a sizeable demand for Farriers.

Related Occupations

Occupations related to the Farrier include: Assistant Horse Trainer, Farm or Ranch Manager, Groom, and Horse Trainer.

FEED AND GRAIN DEALER

Description

Feed and Grain Dealers supply the horse owner with commercial feeds, hay, bedding materials, vitamin and mineral supplements, horse-related equipment, supplies, and medicines. They may act as an authorized dealer representing one particular feed company, or they may act as an independent dealer by providing their customers with a variety of feeds from various feed companies. They usually store and stock their inventory in one retail location and provide delivery service to their customers.

Educational and Training Requirements

Feed and Grain Dealers should have a high school education plus a two- or four-year college degree in Agriculture, Agronomy, Animal Science, or Agribusiness. College-level courses should include marketing, accounting, computer science, and economics. Practical experience at a feed dealership or general farm experience would be desirable. Any practical experience with horses is beneficial but is not mandatory.

Personal Qualifications

The Feed and Grain Dealers must have a high degree of communication skills in dealing with clients and employees on a daily basis. As with other businesspersons, they must be dependable, trustworthy, and reliable. They must be able to work long hours as well as weekends. Most of the work is performed indoors, but on occasion work is performed outdoors in all types of weather conditions.

Licenses and Certification

There is no specific license or certification required for a Feed and Grain Dealer.

Employment Outlook

The employment outlook for the Feed and Grain Dealer is optimistic for the future, since the pleasure horse population in the United States is increasing.

Related Occupations

Occupations related to the Feed and Grain Dealer include: Equine Accountant or Bookkeeper, Retail Tack Shop Operator, and Veterinary Pharmaceutical Salesperson.

FOALING ATTENDANT

Description

The duty of the Foaling Attendant is to be on hand when a mare is ready to give birth. The Attendant must assist the mare in delivery of the foal if necessary. If a veterinarian is called to aid the mare during parturition, the Foaling Attendant must be available to assist the veterinarian as well. Foaling Attendants must become familiar with all the habits and quirks of the broodmares in residence on the breeding farm. They must become familiar with all the signs of parturition that are normally displayed by foaling mares. Foaling Attendants must be able to determine if the parturition procedure (foaling) is normal or if the mare is experiencing complications. Complications would include one of the many abnormal birth positions of the foal, retained afterbirth, abnormal breathing, mare or foal unable to get up, contracted tendons, and so on. After foaling is completed, the Attendant must disinfect the navel stump of the foal with iodine, dry the body of the foal, and make sure the foal is standing and nursing. The Attendant must note whether the foal has passed its first fecal material (meconium) and, if not, must administer an enema to the foal. It is also the responsibility of the Attendant to monitor the foal's body temperature and to observe the newborn foal on a daily basis. If the practice of imprinting foals is performed, the Foaling Attendant must initiate this training procedure immediately after birth.

Educational and Training Requirements

A high school education is a requirement for this position. A two-year degree in Equine Science or Veterinary Technology is favorable but is not essential. Practical experience as a groom or a veterinary assistant would enhance your horsemanship skills and make you more employable.

Personal Qualifications

Foaling Attendants must be in excellent health and physically fit and must have a thorough understanding of basic horsemanship and equine physiology and anatomy. They must be alert and be able to work irregular hours, and they may be required to work independently without supervision. A thorough knowledge of postparturition procedures is necessary, as well as the ability to communicate intelligently to fellow workers, farm managers, and veterinarians. Foaling Attendants may be required to work evenings, holidays, and weekends.

Licenses and Certification

No license or certification is required for the profession of Foaling Attendant.

Employment Outlook

With the reduction in large breeding farms, the employment outlook for the Foaling Attendant will diminish in the future. Certainly a good background in general horsemanship skills would enhance the possibility of employment.

Related Occupations

Occupations related to the Foaling Attendant include: Equine Veterinary Assistant, Farm or Ranch Manager, and Groom.

GROOM

Description

The profession of Groom refers to both the racetrack and the show horse industry. Grooms are responsible for the feeding, watering, and general well-being of the horse. They must place the riding or harness equipment on the horse for the rider or driver. After an exercise or training session for show or racing, the Groom must bathe and blanket the horse and walk it until it is "cooled out." Some Grooms must be able to clip the coat of the horse with electric clippers. They must clean out the stall daily and keep the barn, stable, and tack room clean. They must provide first aid to the horse, and bandage the limbs. The Groom should be able to braid the mane and tail. Finally, the Groom must clean and polish saddles, bridles, harnesses, halters, and other equipment.

People who love horses get satisfaction from working with and helping horses. However, some of the work may be physically demanding and unpleasant. Some duties like euthanizing a hopelessly injured horse or aged animal may be emotionally stressful. Grooms are constantly exposed to bites, kicks, and even diseases that are transmitted to humans from horses. Hours are for the most part irregular. Most full-time workers work 40 to 50 hours per week. Grooms for show and race horses are required to travel to various competitions.

Educational and Training Requirements

Most Grooms are trained on the job. There are some vocational Equine Science programs offered on the high school level. Membership in youth organizations such as 4-H, Future Farmers of America (FFA), and Pony Clubs is helpful in the early stages of education. A high school education is required, as well as practical horse experience in order to properly care for horses. Volunteer work at riding stables or veterinary clinics is certainly recommended for this position. Some summer camps may offer an equitation program whereby practical horse experience may be obtained. Two- or four-year college Equine Science programs may be offered at private and public schools for persons seeking advanced careers as farm/ranch managers or veterinary technicians.

Personal Qualifications

Grooms must be patient, kind, and alert and must possess a thorough understanding of horse behavior. They must be in excellent health and physically fit in order to handle unruly horses. The ability to get along with fellow workers is essential. The Groom should be able to work weekends and holidays. Grooms must be able to relocate and travel to various horse show competitions and racetracks. They are required to work outdoors in all types of weather conditions. Finally, the Groom must have a genuine love for the horse.

Licenses and Certification

The professional Groom on the racetrack must obtain a license in the state in which racing is being conducted. The Groom need only to complete the appropriate forms and pay the required fee in order to obtain a Groom's license. There is no licensing requirement for the show horse Groom.

Employment Outlook

Employment opportunities for the professional Groom generally are expected to be good in the future. Employment is expected to grow; as the population and economy expand and horse ownership increases, more horses will require more caretakers to provide essential care. The turnover is quite high for this profession because it involves hard physical labor and low pay; therefore, the overall availability of grooming jobs is very good.

Related Occupations

Occupations related to the professional Groom include: Exercise Rider, Farrier, Horse Trainer, and Riding Instructor.

GROUNDSKEEPER

Description

Groundskeepers are responsible for the care and maintenance of horse-related facilities. They are responsible for maintaining the grounds of racetracks, fairgrounds, arenas, and polo fields. They make sure supplies and equipment (trees, plants, grass seed, soils, and other materials) are available or on order. They assign various tasks to workers. They direct workers who prepare grounds; apply seeds, fertilizers, and insecticides; and plant trees and shrubs. They plan and oversee irrigation and watering projects. They also

inspect work to be sure it meets certain requirements. Groundskeepers in charge of a large facility like a racetrack usually have a full-time work crew of over 20 workers. They must work within a budget that covers costs for supplies, equipment, and materials.

Educational and Training Requirements

The Groundskeeper should have a high school education. High school level programs may offer courses in landscaping, horticulture, and agriculture. A two- or 4-year college degree is essential for this occupation. College level courses may include horticulture, botany, agronomy, agricultural engineering, entomology, equine science, and landscape design. Business courses may include accounting, advertising, and marketing. Most Groundskeepers obtain on-the-job training from greenhouse firms or landscaping contractors.

Personal Qualifications

Groundskeepers must be able to direct and give instruction to workers. They must enjoy working outdoors and enjoy working with plants, shrubs, and trees. They should be self-motivated and possess good communication skills. They are required to work mostly during daylight hours. They may be required to work more hours during spring, summer, and fall than during the winter months. Additional hours may be required for record-keeping and budget preparation. Since they must work outdoors, they must be able to deal with inclimate weather. Some work may be strenuous and may require the operation of machinery and the use of tools and equipment. Groundskeepers may be required to handle hazardous chemicals and pesticides.

Licenses and Certification

Most states require Groundskeepers who utilize insecticides, pesticides, and fungicides to be certified. Certification requires the passing of a written test on the safe and correct use of insecticides, pesticides, and fungicides. The Professional Grounds Management Society offers certification to Groundskeepers who have a combination of 8 years of experience and a formal education beyond high school, and who successfully complete a course and a test.

Employment Outlook

Employment opportunities are expected to be good in the future. Increased leisure time and interest in equine activities have brought large numbers of people to racetracks, polo fields, country clubs, and fairgrounds. Available positions should be plentiful, especially for those individuals with a college education as well as on-the-job training.

Related Occupations

Occupations related to Groundskeeper include Architect, Equine Fence Dealer or Installer, and Farm or Ranch Manager.

FIGURE 10.10
The Harness Horse Driver is responsible for exercising and racing the standardbred racehorse. (Courtesy of United States Trotting Association, Ohio)

HARNESS HORSE DRIVER

Description

The duty of the Harness Horse Driver is to drive a standardbred racehorse with a racing sulky during a race. In harness racing, it is usually the trainer who drives the horse during a race; however, owners and trainers sometimes hire professional independent drivers to drive their harness horses during a race. Drivers who are not the trainer of the horse they drive are called Catch Drivers. The Catch Driver is an independent driver who drives a horse in a race for a trainer or owner for a fee. The driver or groom generally warms the horse up prior to the race, usually between early races. The Harness Horse Driver checks all equipment before the race, mounts the sulky cart, and guides the horse to its assigned post position behind the car-driven starting barrier. After the race is over, the driver returns the horse to the groom in the paddock. (See Figure 10.10.)

Educational and Training Requirements

A high school education is required with a basic knowledge of standardbred racehorses and harness racing. Initial training may include employment at the racetrack or training farm as a hot walker, groom, stable foreperson, and assistant trainer. As a groom within the harness racing industry, you are not only required to care for the horse but you are responsible for the horse's daily exercise as well. The groom places the harness and cart on the horse and jogs the horse on the racetrack under the supervision of the trainer. As a groom, you learn the skills necessary to prepare you for a career as a Harness Horse Driver.

Personal Qualifications

Harness Horse Drivers do not have to be concerned to the extent a thoroughbred jockey does with their weight. They must have good communication skills when dealing with trainers, owners, and stable employees. They must have a genuine love for the horse as well as the sport of harness racing. The Harness Horse Driver must be willing to work long hours each day and to travel to various harness racetracks throughout the country. Harness Horse Drivers perform most of their work outdoors in all types of weather conditions. Finally, they must have good reflexes, good distance judgment, good eyesight, and common sense.

Licenses and Certification

All Harness Horse Drivers must be licensed by the United States Trotting Association (USTA) in each state in which they compete. Licensing usually requires that the candidate pass a written examination as well as a practical driving test, both of which are issued by the executive vice president of the USTA or the USTA district track committee. A candidate must be sponsored by experienced licensed drivers attesting to the candidate's abilities and experience before a license is issued.

Employment Outlook

There is a general decline in the number of racetracks offering harness racing in this country. Attendance at harness horse racetracks has also declined in recent years due to the increase in off-track betting sites for the betting public. This overall industry decline will lead to a decline in employment opportunities for the Harness Horse Driver. One promising note is that many new racetracks being built in this country are offering mixed racing, that is, standardbreds, thoroughbreds, and quarter horses are all granted racing dates at these racetracks. Some racetracks offer mixed racing on a single day to appease their customers.

Related Occupations

Occupations related to the Harness Horse Driver include: Assistant Horse Trainer, Groom, and Horse Trainer.

HIPPOTHERAPIST

Description

The term *hippotherapist* is not an accurate term, since it does not actually represent a separate occupation. The American Hippotherapy Association (AHA) states that the term implies that hippotherapy is a separate form of therapy instead of what it is, which is the use of the horse's movement as a modality or tool by specifically trained physical therapists, occupational therapists, and speech and language pathologists. Hippotherapy is a treatment approach that utilizes activities on the horse to help patients improve in function, coordination, mobility, and balance. The hippotherapist is a physical or occupational therapist, psychotherapist, or speech therapist with training in hippotherapy through the AHA, a division of the North American Riding for the Handicapped Association (NARHA). Disabilities such as cerebral palsy, arthritis, multiple sclerosis, psychological disorders, and spinal injuries are treatable by Hippotherapy. The Hippotherapist documents the patient's progress, conducts an evaluation, plans treatments, and establishes goals for the patient. (See Figure 10.11.)

FIGURE 10.11
The Hippotherapist utilizes the horse as a therapy tool to allow the patient to improve in function, coordination, mobility, and balance. (Photo courtesy of Terry Brown)

Educational and Training Requirements

A high school education is a requirement for this profession. In addition to a basic high school education, a bachelor's, master's or doctorate degree in Occupational Therapy, Psychotherapy, or Speech Therapy may also be required. College-level courses should include physics, biology, chemistry, human growth and development, biomechanics, evaluation techniques, and therapeutic procedures. Practical experience as a volunteer with a physical therapist, psychotherapist, or speech therapist in a hospital or clinic is recommended for this profession. A Hippotherapist should be a skilled horse person able to demonstrate riding skills and to understand the movement of the horse. One might volunteer as a side walker to a professional Hippotherapist to gain valuable experience in this field.

Personal Qualifications

Hippotherapists must have strong communication skills in order to help their patients understand the treatments. They must also be compassionate and possess a strong desire to assist patients in adjusting to overcoming their disabilities. Hippotherapists must be able to understand the behavior of horses in order to choose the best horse for the patient, and they must have physical stamina and manual dexterity when dealing with both patients and horses. They must be able to lift, kneel, crouch, stand, and walk for long periods of time. Work is performed either outdoors in a riding ring or indoors in an indoor riding arena.

Licenses and Certification

A Hippotherapist is actually a physical or occupational therapist, psychotherapist, or speech therapist. All states require these professionals to pass a written examination upon graduation from an accredited educational program before they practice within their respective fields. A certification examination in the clinical specialty of hippotherapy is held twice annually by the AHA for licensed physical therapists, occupational therapists, and speech therapists who meet certain criteria. The criteria includes:

- license or credential as a physical therapist, occupational therapist, or speech therapist
- three years of full-time clinical practice in physical therapy, occupational therapy, or speech therapy
- at least 100 hours of hippotherapy practice within the three years prior to certification
- independent riding ability

Some states require these professionals to participate in continuing education courses in order to meet licensing requirements.

Employment Outlook

For the most part, the future of this profession appears to be promising. Hippotherapy is becoming more acceptable within the realm of health insurance reimbursement, and this area is also growing because this type of therapy can be enjoyable for the patient. It takes the patient out of the clinical setting into a nonclinical setting in order to achieve treatment goals.

Related Occupations

Occupations related to the Hippotherapist include: Equine Chiropractor, Equine Massage Therapist, and Riding Instructor.

HORSE BREEDER

Description

The Horse Breeder breeds and raises horses in order to improve a particular breed, to develop new breeds, and to maintain the standards of existing breeds. Horse Breeders strive to produce horses with the characteristics and traits that suit the purpose of the individual owners. They mate stallions and broodmares with special traits in order to produce offspring with the best traits of both parents. Horse Breeders are responsible for the well-being of the horses until they are sold privately or at public auction. Most of their work is performed outdoors and inside barns, breeding sheds, and laboratories. Hours are often irregular, as the estrus (heat) cycles of brood-mares vary greatly. The Horse Breeder may only work for few hours a day during the nonbreeding season and up to 24 hours per day during breeding season.

Educational and Training Requirements

A high school education is essential for the Horse Breeder. The Horse Breeder must have a thorough knowledge of the horse's reproductive system as well as the principles of genetics. Most professional Horse Breeders have a bachelor of science degree in Animal Husbandry, and most college programs in Animal Husbandry include courses in animal behavior, equine nutrition, chemistry, zoology, genetics, business management, biology, and biochemistry. Many colleges and universities offer master's and doctorate degrees in Animal Reproduction and Anatomy and Physiology. High school students may prepare for a career as a Horse Breeder by taking courses in biology, life sciences, mathematics, computer sciences, and business.

Personal Qualifications

Horse Breeders must have good horsemanship skills. They must be able to make decisions, supervise employees, and think clearly in emergency situations. They must be alert, swift, and agile in order to avoid the dangers of a biting or kicking horse. The Horse Breeder must be able to work long, irregular hours, including evenings, weekends, and holidays. Traveling to horse shows, breeding seminars, racetracks, and other breeding farms is essential. Finally, Horse Breeders must have good communication skills, both oral and written, when dealing with veterinarians, owners, trainers, employees, and clerical staff.

Licenses and Certification

There is no licensing requirement for the Horse Breeder, but certain breed registry rules and regulations must be followed. For example, the Jockey Club governing all thoroughbred registrations strictly prohibits the practice of artificial insemination as a method of breeding thoroughbreds.

Employment Outlook

The financial and employment outlook for the Horse Breeder is uncertain. The economic trends in the horse industry are based on supply and demand as well as the state of the nation's economy. Success as a Horse Breeder will vary from breed to breed.

Related Occupations

Occupations related to the Horse Breeder include: Bloodstock Agent, Breeding Technician, Equine Geneticist, and Farm or Ranch Manager.

HORSE INDUSTRY SECRETARY

Description

Breeding farms, veterinary hospitals, as well as training and riding stables employ the Secretary to perform and coordinate office activities. Secretaries are responsible for relaying information to both staff members and clients. They are also responsible for clerical and administrative duties that are required to operate and maintain a horse facility efficiently. They answer telephone calls, schedule appointments, maintain files, conduct tours, sort mail, and type letters. The Horse Industry Secretary must be well versed in operating all types of office equipment, such as telephone paging systems, postal machines, facsimile machines, photocopiers, and personal computers.

Educational and Training Requirements

A high school education is required with a basic knowledge of office skills. Secretaries must have good language arts skills, including spelling, grammar, punctuation, and public speaking. Computer skills are necessary for this profession. Any knowledge of or practical experience with horses or horsemanship is beneficial. Training may be acquired at community colleges, business schools, and vocational schools offering one- or two-year programs in Secretarial Sciences.

Personal Qualifications

Secretaries employed within the horse industry usually work indoors in an office setting. They are required to type or view a video display monitor while sitting for long periods of time. They must be tactful, discrete, well organized, and have good judgment. They must have good communication skills, both written and oral, when dealing with employers, clients, and fellow employees.

Licenses and Certification

There is no formal licensing requirement for the Horse Industry Secretary. However, a Secretary with considerable experience may earn the designation of Certified Professional Secretary (CPS). This title is earned by passing a series of examinations.

Employment Outlook

Employment opportunities for the Horse Industry Secretary should be favorable in the near future, especially for well-qualified and experienced Secretaries with a background in basic horsemanship. The major influence in limiting employment growth for most Secretaries is the increased development and use of office technology.

Related Occupations

Occupations related to the Horse Industry Secretary include: Equine Accountant or Bookkeeper, Equine Advertising Specialist, Equine Veterinary Assistant, and Horse Insurance Agent or Broker.

HORSE INSURANCE AGENT OR BROKER

Description

The duty of the Horse Insurance Agent or Broker is to sell insurance policies to protect the horse owner from financial loss. Insurance policies protect horse owners against death, injury, illness, and liability involving their horses. Horse Insurance Agents are employed by a single horse insurance company; they sell and service insurance policies offered only by their employers. Horse Insurance Brokers, on the other hand, are not employed by a specific horse insurance company, but represent several horse insurance companies. They sell and service insurance policies for their clients with the company that offers the best rate and coverage. The Agent simply works for the insurance company, while the Broker simply works for the client. A wide variety of insurance policies is available to the horse owner, such as mortality, loss of use, limited perils, transportation, stallion fertility, unborn foal insurance, care, custody and control, medical expenses, liability, and workers' compensation.

Educational and Training Requirements

A high school education as well as a two- or four-year college degree in Business is recommended for this position. College courses should include economics, marketing, accounting, and computer science. Practical experience in the field of insurance while employed at an agency or home office of an insurance company is beneficial in gaining knowledge for this profession. A general knowledge of horses is helpful but is not a requirement.

Personal Qualifications

Horse Insurance Agents or Brokers must enjoy meeting and working with people. They must have an appreciation for and general knowledge of the horse industry. They must also have good communication skills, both written and oral, and they must be outgoing, confident, and persuasive. Most companies and clients expect Insurance Agents and Brokers to be neat, courteous, and dependable. They must be able to work independently with little or no supervision. Most of their work is performed indoors within an office setting; however, there are occasions when the Horse Insurance Agent or Broker must work outside the office setting when meeting with clients, prospects, and associates. The Horse Insurance Agent usually works for a set salary, while the horse Insurance Broker usually receives a set commission percentage or a commission plus a monetary bonus on all insurance policies sold.

Licenses or Certification

All states require Insurance Agents and Brokers to be licensed. In order to be licensed, the applicant must successfully pass a written examination based on state insurance laws. Some horse insurance companies conduct classes to prepare their prospective Agents for the licensing examination. Horse Insurance Agents and Brokers may also earn the Chartered Property Casualty Underwriter (CPCU) title by passing several written examinations offered by the American Institute for Property and Liability Underwriters.

Employment Outlook

In general, the insurance industry, including the horse insurance industry, is expected to grow in the future. Horse owners need to purchase insurance to cover the increasing values of their horses, property, and liability; therefore, the demand for the Horse Insurance Agent or Broker is expected to grow.

Related Occupations

Occupations related to the Horse Insurance Agent or Broker include: Appraiser, Bloodstock Agent, Equine Attorney, Horse Insurance Claims Adjuster, and Horse Insurance Underwriter.

HORSE INSURANCE CLAIMS ADJUSTER

Description

Horse Insurance Claims Adjusters deal with a variety of problems concerning claims and at the same time maintain good relations with policyholders. They handle complaints and interpretation of the policy conditions. When a claim is filed for property damage or for loss of a horse due to injury or death, the Claims Adjuster determines whether the client's insurance policy covers the loss and the amount of the loss. Claims Adjusters plan and schedule the work required to process claims. They investigate claims by interviewing policyholders, witnesses, and veterinarians and by inspecting the damaged property to determine the extent of the company's liability. They are required to prepare reports of their investigation. The Adjuster also negotiates with the policyholder in order to settle the claim. When a claim is contested, the Adjuster may be required to testify in court.

Educational and Training Requirements

A high school education is essential. A two- or four-year degree is not required but will enhance the possibility of employment. While a college education is preferred, it is not essential, as employment may be based on related work experience. Courses in computer science are most helpful for this position, and any practical insurance experience is beneficial. Horse experience is beneficial but is not required.

Personal Qualifications

The Horse Insurance Claims Adjuster must have good communication skills when dealing with policyholders, witnesses, and other professionals in resolving a claim. The work of Claims Adjusters is performed mostly indoors in an office setting, although on occasion they may be required to travel to obtain information through personal interviews. They must have a pleasant personality when dealing with irate policyholders. Most of the work is routine and requires sitting at a desk and a computer terminal for long periods of time. Interview schedules may include evenings and weekends, depending on the availability of the parties involved.

Licenses and Certification

Licensing is mandatory in most states for the Claims Adjuster. The applicant must pass a written examination, administered by the state Insurance Department, before a license is issued. Most training for this position is usually received in the home office of the insurance company.

Employment Outlook

With the present increase in horse ownership, more and more horse owners protect their financial investment through insurance. Because much of the Claims Adjuster's work is centered on direct contact with clients, the demand for Horse Insurance Claims Adjusters is expected to grow with the increase in the number of clients.

Related Occupations

Occupations related to the professional Horse Insurance Claims Adjuster include: Appraiser, Bloodstock Agent, Equine Accountant or Bookkeeper, Equine Attorney, Horse Insurance Agent or Broker, and Horse Insurance Underwriter.

HORSE INSURANCE UNDERWRITER

Description

The duty of the Horse Insurance underwriter is to evaluate and select the horse-related risks the insurance company will insure. Horse Insurance Underwriters decide whether or not applicants for horse-related insurance are an acceptable risk. They evaluate the information detailed on the application form for insurance, including veterinary medical reports. They decide whether to issue a policy and determine the amount of premium. Horse Insurance Underwriters must correspond with Insurance Brokers and Agents and, in some cases, policyholders. They are usually employed by the insurance company at the home office and are considered salaried employees.

Educational and Training Requirements

A high school education is essential as well as a two- or four-year degree in Business, Accounting, Business Law, or Economics. Entry-level positions include assistant underwriters and underwriter trainees.

Personal Qualifications

Horse Insurance Underwriters must possess good judgment in order to make sound decisions. They must have good communication skills when dealing with insurance brokers, agents, and policyholders. They must also be aggressive when obtaining information for a particular risk from outside sources. All work is performed indoors in an office setting. Travel is limited, and the position does not require any unusual physical activity.

Licenses and Certification

There is no mandatory licensing requirement for the Horse Insurance Underwriter. However, Underwriters may earn the more advanced Charted Property Casualty Underwriter (CPCU) title, which generally takes about five years and requires the passing of 10 examinations administered by the American Institute for Charted Property Casualty Underwriters.

Employment Outlook

The employment outlook for the Horse Insurance Underwriter is expected to be good in the near future. Most underwriting positions are expected to result from the need to replace underwriters who transfer to other occupations or retire.

Related Occupations

Occupations related to the Horse Insurance underwriter include: Appraiser, Bloodstock Agent, Equine Accountant or Bookkeeper, Equine Attorney, Horse Insurance Agent or Broker, and Horse Insurance Claims Adjuster.

HORSE SHOW AND EVENT JUDGE

Description

The responsibility of the Horse Show and Event Judge is to determine the placement of each individual class at a horse show or event. The Judge must enforce the rules and regulations pertaining to each specific show or event. The Judge is usually hired by the chairperson of each show or event. Judges may represent various breeds of horses, breed associations, and horse-related organizations at shows and events. They are hired for different equine disciplines such as dressage, jumpers, rodeos, and English and Western equitation. The Horse Show and Event Judge may be required to judge horses on their conformation as well as their performance.

Educational and Training Requirements

A Horse Show and Event Judge should have a high school education with some vocational training in general horsemanship. A two- or four-year degree in Equine Science or Animal Science is also beneficial for this profession. Courses should include anatomy and physiology, conformation, and equitation. Practical experience at horse shows and events is an asset to the aspiring Judge.

Personal Qualifications

The Horse Show and Event Judge must have a pleasant attitude and personality. Good communication skills are important when dealing with officials and contestants. The Judge must be a well-organized individual and must be observant and fair. Judges must be willing to work long hours, including weekends. Most of their work is performed outdoors where they are subject to all types of weather conditions. If work is performed indoors, it usually takes place within an indoor riding ring. Judges must be able to make decisions and to stand by their decisions, and they should be able to direct people and operations and possess the foresight to carry out long-term goals.

Licenses and Certification

Horse Show and Event Judges may be required to become certified by the specific organizations they represent at particular shows. For example, if the American Quarter Horse Association were to conduct a horse show for registered quarter horses, a Judge certified by the American Quarter Horse Association would be utilized. Certification may require candidates to pass a written test covering the goals, rules, and regulations of the horse association or organization they will represent. Candidates for judgeship may also be required to act as assistant judges or ring stewards at a recognized show to gain some practical experience. The candidate for a judgeship should have successfully competed in horse shows and events in various disciplines of riding and showing.

Employment Outlook

The need for the professional Horse Show and Event Judge should increase in the near future. Pleasure and show horse ownership are on the rise in the United States, and with this increase in ownership, there is an increase in the number of horse shows and events each year. A growing population and more disposable income will result in the purchase of more horses and an increase in horse-related leisure activities. This will translate into demand and employment for the professional Horse Show and Event Judge.

Related Occupations

Occupations related to the Horse Show and Event Judge include: Breeding Technician, Horse Trainer, and Riding Instructor.

HORSE TRAILER AND VAN DEALER

Description

Horse Trailer and Van Dealers sell, service, and repair horse trailers and horse vans, and they supply their customers with parts and equipment. Horse trailers are designed to be pulled by a motor vehicle; horse vans do not have to be pulled. A horse van is one unit consisting of a vehicle with a compartment behind the cab designed to hold the horse safe and secure during transport. Some Dealers sell and service a broad range of trailers and vans for several different manufacturers, while others sell and service only certain trailers and vans for one or two specific manufacturers. Horse Trailer and Van Dealers purchase and stock horse trailers and vans as well as related equipment that they think will sell. They set the prices for the trailers and vans as well as their services in order to cover business costs and make a profit. They determine the allowance for trade-ins from customers. Dealers must supervise all office workers and are responsible for the hiring and training of all sales and service staff. They usually service and repair the trailers and vans they sell and also stock and sell replacement parts for their products.

Educational and Training Requirements

A high school education is essential. A two- or four-year college degree in Business is helpful. College courses should include business management, marketing, economics, computer science, and accounting. Course in equine science are also beneficial. A general knowledge of horses and horsemanship is not required but will aid Dealers in talking to and understanding the needs of their customers. Any practical experience and training at a trailer and van dealership are also beneficial for this position. Some Dealers begin as a staff member of the service department and then move up to a position within the parts and sales department. In order to become a Horse Trailer and Van Dealer, they must secure a contract with one or more trailer or van manufacturers.

Personal Qualifications

The Horse Trailer and Van Dealer must be willing to work long, irregular hours including weekends and holidays. Work is performed both indoors and outdoors, in an office setting, a showroom, and a sales display lot. Dealers must be friendly, patient, persistent, honest, and discreet. They must have good communication skills and administrative and technical abilities. Finally, Dealers must have a thorough knowledge of horse trailers and vans and the ability to sell and service them. They must also have some knowledge of horses in order to understand the needs of their customers.

Licenses and Certification

There is no mandatory licensing or certification requirement for the Horse Trailer and Van Dealer.

Employment Outlook

The need for horse trailers and vans should increase in the near future. Horse ownership and population are on the rise in the United States; with this increase, there will be an increase in the number of horse shows and events each year. This will translate into a demand for horse trailers and vans in order to transport horses to horse shows and events.

Related Occupations

Occupations related to the professional Horse Trailer and Van Dealer include: Farm Equipment Dealer, Feed and Grain Dealer, and Retail Tack Shop Operator.

HORSE TRAINER

Description

The Horse Trainer trains horses in various disciplines such as racing, dressage, reining, polo, cutting, and jumping. The basic duty of the Horse Trainer is to assess the potential of the horses placed in his or her care. Horse Trainers may be involved in teaching horses gaits and working toward building horses to their full potential. Horse Trainers must also supervise employees who exercise, feed, and provide general care for the horses, and they are required to assist horse owners in buying and selling horses. Most Horse Trainers have a farm, ranch, or stalls at a racetrack or training center where they apply their trade and offer their services. When horses are sent to a Horse Trainer by an owner, it is usually for a monthly boarding or training fee. The Horse Trainer must provide the feed, bedding, labor, shipping, veterinary care, and farrier work.

The Racehorse Trainer is required to set up a training program for each individual horse. It is the job of the Trainer to prepare the horse for racing by allowing the horse to become accustomed to the starting gate and the racing and training equipment. The Trainer prepares the horse with exercise so it will reach the peak of its performance. The Racehorse Trainer supervises the grooms, exercise riders, jockeys, drivers, and others who provide the general care of the horse. Racehorse Trainers are responsible for entering and scratching a horse to and from races. They must be in contact with the owners of the horses in their care on a regular basis to keep them informed, and they must deal with insurance matters concerning their operation, such as workers' compensation insurance. The working hours and conditions vary with the type of horses being trained. Factors affecting the hours include:

◆ amount of travel time
◆ number of horses in training
◆ number of employees
◆ type of training

For the most part, hours are irregular. For example, the Racehorse Trainer usually starts at daybreak and works for 4 to 5 hours. If a horse is scheduled to race on a particular day, the hours are much longer. Income for a Racehorse Trainer is based on a daily rate ranging from $25 to $100 per day, plus a standard 10 percent of the horse's earnings. Other types of Horse Trainers who operate their own business charge their clients a monthly rate that varies with the experience of the Trainer and the location of the operation. (See Figure 10.12.)

Educational and Training Requirements

For this profession, a high school education is required. In addition to a basic high school education, practical experience in areas of interest, such as racing, polo, dressage, and so on, may prove beneficial for this profession. A two- or four-year degree in Equine Science, Business, or Computer Science is also recommended for this profession.

Personal Qualifications

The Horse Trainer should be in excellent health and be willing to work long hours each day. Horse Trainers must have a genuine love for the horse. They must have good horsemanship skills

FIGURE 10.12
The Horse Trainer must provide for the daily care and training of the horse. (Photo courtesy of Michael Dzaman)

and yet be gentle, firm, and patient when training horses. They must have a great deal of respect for the horses they train and possess a basic knowledge of horse behavior. Knowledge of nutrition and conditioning is essential for this profession. Trainers must also have good communication skills, both written and oral. Because of the increasing number of Hispanic laborers in the horse industry, it is most helpful for the Horse Trainer to speak and understand the Spanish language.

Licenses and Certification

Racehorse Trainers must be licensed in each state or racetrack in which they operate their business by meeting the following requirements:

1. possessing at least three years' experience on the racetrack as a licensed groom, assistant trainer, exercise rider, and so on
2. passing a written examination given by the state racing commission at a particular racetrack
3. passing a practical examination by placing bandages on a horse, saddling and bridling a horse, and answering questions correctly concerning equipment
4. after receiving a license:
   ◆ providing the state racing commission with evidence of a workers' compensation insurance policy
   ◆ providing the state racing commission with a list of horses under the Trainer's care as well as a list of their owners

Other types of Horse Trainers may become members of local or national horse organizations, such as the Professional Horseman's Association.

Employment Outlook

Employment for the professional Horse Trainer is limited in scope and depends a great deal on the overall economy. At present, the horse population in the United States is growing and there is a need for the professional Horse Trainer for show and pleasure horses. In contrast, the prospects for the Racehorse Trainer are very limited because of the recent decline in the number of racetracks in the country offering pari-mutuel wagering. As stall space diminishes, so does the need for new Trainers. However, opportunity for the Racehorse Trainer does exist on large breeding farms and training centers throughout the country.

Related Occupations

Occupations related to the Horse Trainer include: Equine Veterinary Assistant, Exercise Rider, Farm or Ranch Manager, Groom, and Jockey.

HORSE VAN OR TRUCK DRIVER

Description

The duty of the Horse Van or Truck Driver is to operate large horse vans and livestock transporting vehicles. Horse Van or Truck Drivers pick up horses and drive them safely to their destinations, including horse shows, racetracks, rodeos, breeding farms, training centers, airports, and veterinary facilities. Their travels may take them on local trips as well as long-distance trips. Horse Van or Truck Drivers are responsible for maintaining their vehicles as well as providing a safe and nonstressful trip for the horses. They must supervise the overall loading and unloading procedures and deal with any emergencies that may occur during transit. Some transporting companies utilize two drivers on very long trips; one drives while the other sleeps. These long trips may last for days, with the van or truck stopping for fuel, food, loading, and unloading. (See Figure 10.13.)

Educational and Training Requirements

A high school education is required as well as practical horse experience in order to properly handle horses during transit. Completion of a driver training program is desirable as a means of preparing for van or truck driving jobs and obtaining a commercial driver's license.

FIGURE 10.13
The Horse Van or Truck Driver is responsible for picking up horses and driving them safely to their destinations.

Personal Qualifications

The Horse Van or Truck Driver must be willing and able to drive for long periods of time and in bad weather conditions, heavy traffic, and unusual road conditions. Some drivers must spend a great deal of time away from home, friends, and family. Drivers on long trips must be able to deal with boredom, loneliness, and fatigue. Although most drivers work during the day, there are many who must drive at night and on weekends and holidays to avoid traffic delays and to pick up and deliver horses on schedule. As drivers must deal with horse owners, dispatchers, and fellow drivers, they must be able to get along with people. They must be responsible, self-motivated individuals, since most drivers work with little or no supervision. A driver must be able to speak well and must exhibit a neat appearance.

Licenses and Certification

All drivers of trucks designed to carry at least 26,000 pounds, which includes most horse vans and tractor trailers, are required to obtain a special commercial driver's license (CDL) from the state in which they legally reside. A driver must be at least 21 years of age and pass a physical examination. Good hearing, 20/40 vision, normal use of arms and legs, and normal blood pressure

are the main physical requirements. In addition, all drivers must take and pass a written examination on the Motor Carrier Safety Regulations of the United States Department of Transportation. Drivers are required to be tested for alcohol and drug use by their employers as a condition of employment and are subject to periodic random tests while working.

Employment Outlook

The employment outlook for the Horse Van or Truck Driver is favorable for persons who are interested in this horse-related profession. Van and Truck Driver positions vary greatly in terms of hours, earnings, actual driving time on the road, and the type of equipment operated. Competition is anticipated for those jobs with the most attractive earnings and working conditions.

Related Occupations

Occupations related to the Horse Van or Truck Driver include: Horse Trailer and Van Dealer.

HOT WALKER

Description

The basic function of the Hot Walker is to walk hot racehorses after a workout or race. The purpose of walking the hot horse is to bring the horse's body temperature down to normal after exercise or a race. Walking is an excellent means of providing daily exercise for the racehorse. Walking may take place either outside the barn, in a walking ring, or under the shedrow of the racing stable. Hot Walkers must be able to regulate the horse's water intake while walking. They should also be able to apply and remove blankets during the cooling-out period. In addition to actual walking duties, Hot Walkers must be able to restrain and hold the horse for the veterinarian, the farrier, or merely while the horse is grazing on grass. They must also be able to hold and control the horse while soaking its limbs in a tub of water or ice and apply various types of therapeutic machines to the limbs and body of the horse. Finally, the Hot Walker is responsible for sweeping, raking, folding blankets, and keeping the shedrow of the barn and tack room neat and clean. (See Figure 10.14.)

Educational and Training Requirements

A high school education is essential for the Hot Walker as well as a general knowledge of basic horsemanship skills. The position of Hot Walker is considered an entry-level position in the racing industry.

Personal Qualifications

Hot Walkers should possess a genuine love for horses. They should be physically fit and alert at all times. The ability to communicate with fellow workers and supervisors is essential. Finally, the Hot Walker should have a calm and pleasant personality when dealing with both people and horses.

Licenses and Certification

All Hot Walkers must be licensed at the racetrack where they are employed. The employer (trainer) must submit the name of the Hot Walker to the licensing department at the racetrack. Licensing

in most states consists of the payment of the required fee, fingerprinting, and photographs taken before an official license (badge) is issued. The license must then be worn at all times on the outer clothing of the Hot Walker while working the stable area of the racetrack.

Employment Outlook

The employment outlook for the Hot Walker appears poor, since more and more racetracks are closing because of financial difficulties. Also, there is an increase in the use of automatic hot walking machines by trainers, thus reducing the need for the services of a Hot Walker. These machines reduce staff and expenses as well as time, since four horses can be walked at one time. Although these hot walking machines are efficient, someone must constantly observe the horses on the machine and regulate the speed. It is in this position where the Hot Walker is still of value.

Related Occupations

Occupations related to the Hot Walker include: Exercise Rider and Groom.

IDENTIFIER

Description

The Identifier is usually employed by a racetrack or racing association. The responsibility of the Identifier is to verify the identity of each horse stabled on the racetrack grounds. Identifiers must also verify the identity of all horses shipping into the racetrack for a particular race. Official Identifiers are responsible for the safekeeping of the foal certificates of all horses stabled on the grounds, and they must know the appropriate trainer and barn number where each horse is stabled. On each racing day, the Identifier pulls the foal certificates and photographs of each horse entered in a race on that day. The Identifier personally inspects each horse in each race by verifying the lip tattoo, body color, head and

FIGURE 10.14
The Hot Walker walks the hot racehorse after a race or exercise.

leg markings, scars, and chestnuts (night eyes). Not all Identifiers are employed at racetracks; some are employed by private companies specializing in horse identification to protect horses from theft. This type of Identifier is involved in freeze-branding the skin under the mane with a coded series of symbols for identification purposes. Others are involved in injecting an electronic microchip under the skin of the horse for positive and unalterable identification.

Educational and Training Requirements

A high school education is a requirement for this position, as well as a two- or four-year college degree in Equine Science or Animal Science. Identifiers should have a thorough knowledge of equine anatomy and physiology. They must also have some practical experience in racetrack operations. Good horsemanship skills are beneficial as well. Backstretch employment as a groom, trainer, exercise rider, and so on would be beneficial.

Personal Qualifications

Identifiers must have a pleasant, polite attitude and be able to communicate intelligently to racing officials, veterinarians, and trainers on a daily basis. The Identifier must be well organized and must be able to keep accurate records on horses. Computer literacy is beneficial to this occupation. Identifiers must be alert at all times in order to prevent ringers. Finally, since most of their work is performed outdoors, they are subject to all types of weather conditions.

Licenses and Certification

No mandatory license or certification is required for the profession of an Identifier.

Employment Outlook

The number of Identifier positions are limited to the number of racetracks in operation. The Identifier is employed at racetracks offering both flat racing with thoroughbreds and quarter horses and harness racing with standardbreds.

County and state fairs offering racing also require the services of an Identifier. With the decline of horse racing in the United States, the need for the services of an Identifier is also in decline. However, those entering careers with private identification and security companies will have greater opportunities for employment in the future.

Related Occupations

Occupations related to the Identifier include: Equine Veterinary Assistant, Groom, and Horse Trainer.

JOCKEY

Description

It is the responsibility of the Jockey to ride the racehorse in a race. The owner and trainer usually select which Jockey to employ for the race. The Jockey mounts the racehorse in the saddling paddock and is given instructions by the trainer as to the manner in which he or she is to ride the horse. The Jockey warms up the racehorse and enters the starting gate prior to the race. Once the race is over, the Jockey returns the racehorse to the unsaddling enclosure. After the race, the Jockey usually relays to the trainer any important information that will help the horse perform in the future. The Jockey is responsible for all riding equipment, such as saddle, elastic girths, helmet, goggles, boots, whip, and riding pants. The owner of the horse supplies the Jockey with the shirt or "silks" to be worn during the race to identify the owner.

Jockeys receive 10 percent of all purse money earned by the horses they ride. In addition to this fee, Jockeys also receive a set fee for each horse ridden in a race regardless of finish. Some Jockeys hire an agent to book or solicit mounts from trainers for upcoming races. For their services, agents usually receive 25 to 30 percent of all purse money earned by the Jockey. On a national basis, Jockeys earn an average of $30,000 per year.

Educational and Training Requirements

A high school education is essential for the Jockey. The Jockey must have basic riding skills as well as practical experience in all phases of horsemanship. Experience as an exercise rider at a racetrack, farm, or training center is beneficial. Initial training may also include employment at the racetrack as a hot walker, groom, stable foreperson, and assistant trainer.

Personal Qualifications

Jockeys should naturally weigh between 100 to 115 pounds. They must be willing to start working early in the morning and to work long hours. They must be willing to ride in all types of weather conditions. Jockeys must be physically fit in order to control 1,000-pound racehorses moving at 35 to 40 miles per hour. They must be willing to travel to ride horses at different racetracks throughout the country. A Jockey must have the ability to think fast and make quick decisions in order to keep a horse out of trouble during the running of a race.

Licenses and Certification

In most states that conduct horse racing, a prospective rider must apply for an Apprenticeship Jockey license. The applicant must meet certain requirements before a license is issued. In some states, an applicant must be at least 18 years of age and have been employed and licensed as an exercise rider at a racetrack for a period of one year. Finally, the prospective Jockey must ride before the official stewards of the racetrack to demonstrate an ability to ride as well as knowledge of the rules of racing.

Upon receipt of an Apprentice Jockey license, a weight allowance, usually referred to as the bug, is granted to the Apprentice Jockey. The bug gives the Apprentice Jockey an allowance of 5 to 10 pounds as an advantage over the rest of the Jockeys at a racetrack. In most racing states, Apprentice Jockeys are allowed 10 pounds until they win five races. The weight allowance is then dropped to 7 pounds until they win an additional 30 races for a total of 35 wins. After 35 wins, Apprentice Jockeys who have not completed their first year of riding since being licensed are given an allowance of 3 pounds. When the first year is completed, they officially become full-fledged Journeymen or Journeywomen.

Employment Outlook

Horse racing occurs throughout the United States and Canada. County and state fairs across the country also offer horse racing to the general public. However, the general decline in the number of racetracks and race horses in the United States has resulted in less employment opportunity for a Jockey. Overall, riding races is a very competitive yet rewarding occupation.

Related Occupations

Occupations related to the Jockey include: Exercise Rider, Harness Horse Driver, Horse Trainer, and Outrider.

JOCKEY AGENT

Description

The function of the Jockey Agent is to secure mounts for the jockey. The Agent is hired by the individual jockey for this task. A good Jockey Agent is familiar with all the trainers on the racetrack grounds as well as their horses. Agents must determine which horses present the best chances for their jockey clients to win. Each racing jurisdiction has rules and regulations governing the duties of the Jockey Agent. Most states allow Agents to handle the bookings of two seasoned jockeys or two apprentice jockeys or one of each category. For obtaining riding commitments for their clients, Agents receive a percentage (usually 25 to 30 percent) of their Jockeys' earnings.

Educational and Training Requirements

A high school education is essential for the Jockey Agent. A two-year college degree in

Business Management is beneficial but is not required. Jockey Agents must have thorough knowledge of the horse racing industry, and they must have the ability to handicap a race in order to determine which horse will have the best chance to win. They must also be familiar with the rules of racing that apply to each racetrack and each state in which they ply their trade.

Personal Qualifications

Jockey Agents must be polite and have a pleasant personality when dealing with owners, trainers, and racing officials. They must be diplomatic and fair in their dealings with trainers. Jockey Agents must have good communication skills when dealing with clients over the telephone and in person. They must be willing to rise early to meet with trainers and to coordinate workouts with their jockeys. Attending races in the afternoon or evening is a must. Finally, Agents must be able to work 7 days per week, tolerate all types of weather, and be willing to work hard, long hours each day.

Licenses and Certification

The Jockey Agent must be licensed by the racetrack and the state racing commission. Before Agents can begin operating, they must submit to a complete background check that includes fingerprints and photographs.

Employment Outlook

The career of a Jockey Agent is very competitive. If jockeys are not satisfied with an Agent's services, they may terminate the Agent. As with all racetrack occupations, the present decline in the number of racetracks and the decline in the number of race horses have led to a decline in the number of jockeys, resulting in a decline in the need for the Jockey Agent.

Related Occupations

Occupations related to the Jockey Agent include Assistant Horse Trainer, Horse Trainer, and Jockey.

JOCKEY VALET

Description

Jockey Valets provide a service to the jockey by laying out the proper shirts (silks) for the jockey for each race, as well as the jockey's helmet, riding pants, safety vest, and goggles. They must keep the jockey's riding boots clean and polished, and they clean and maintain the jockey's saddle and girths. One of the most important duties of the Jockey Valet is to assist the trainer in tacking and untacking the racehorse before and after each race. The Jockey Valet is usually employed by the racetrack to provide this service to the jockey.

Educational and Training Requirements

A basic high school education is all that is required, along with a good background in the racing industry with some practical experience on the backstretch in various capacities. The Jockey Valet must be knowledgeable about racehorse equipment and must be able to properly place a racing saddle on a racehorse.

Personal Qualifications

Jockey Valets must be willing to work long hours and to work outdoors in all types of weather conditions. They must be patient and knowledgeable about equine behavior, especially when dealing with unruly horses during the saddling procedure. They must have a pleasant personality in order to deal with jockeys, owners, trainers, and racetrack officials.

Licenses and Certification

There is no specific license or certification required for a Jockey Valet. The only license requirements are those for all racetrack employees, which entail fingerprinting, photographs, and background checks.

Employment Outlook

Employment as a Jockey Valet is limited to the number of racetracks in existence. As with any racetrack occupation, the opportunity for employment is declining with the overall decline of racing facilities in the country.

Related Occupations

Occupations related to the Jockey Valet include: Assistant Horse Trainer, Exercise Rider, Groom, Horse Trainer, Jockey, and Pony Boy or Girl.

LABORATORY ANIMAL TECHNICIAN

Description

Laboratory Animal Technicians process, examine, and analyze samples of body fluids, tissues, and cells obtained from horses. They look for parasites, bacteria, viruses, and other microorganisms. They also perform blood tests and analyze the chemical content of body fluids. Laboratory Technicians may be involved in testing for drug levels in horses from samples for examination, counting cells, and searching for abnormal cells. In addition to microscopes and cell counters, they utilize sophisticated technical equipment and instruments that perform a series of tests at the same time. They then analyze the results of the tests and relay their findings to the equine veterinary practitioners. Laboratory Technicians employed in a small laboratory may perform many different types of tests, while those employed in a large laboratory will generally specialize in one particular testing procedure. (See Figure 10.15.)

Educational and Training Requirements

Laboratory Technicians generally have a two-year college degree or a certificate from a vocational or technical school, a hospital, or the Armed Forces. Four-year degree programs in Medical Technology should include courses in chemistry, biological sciences, microbiology, and mathematics. Other college programs available include specialized courses focused on the knowledge and skills used in the clinical labora-

FIGURE 10.15
The Laboratory Animal Technician analyzes and examines body samples obtained from horses. (Courtesy of Agricultural Research Service, USDA)

tory. Programs may also offer or require courses in management, business, and computer skills. Master's degree programs in Medical Technology Sciences provide training in specialized areas of laboratory work including teaching, research, and administrative duties. Very few people become technicians solely through on-the-job training. Due to the use of sophisticated equipment and instruments, computer skills are essential for this profession.

Personal Qualifications

The Laboratory Technician requires analytical judgment and the ability to work under pressure. The ability to pay close attention to details is also a required trait; manual dexterity and normal color vision are desirable traits as well. Laboratory Technicians must be very good at problem solving, have mechanical aptitude, and be able to follow detailed instructions. Finally, they must be patient, reliable, and possess a pleasant personality when dealing with superiors and fellow workers.

Licenses and Certification

Some states require laboratory personnel to be licensed or registered. Information on licensing is available from state departments of health, boards of occupational licensing, and occupational information coordinating committees. Certification is a prerequisite for most positions and often is required for advancement. Agencies that certify Laboratory Animal Technicians include the Board of Registry of the American Society of Clinical Pathologists, the American Medical Technologist, the National Certification Agency for Medical Laboratory Technology, and the American Association for Laboratory Animal Science (AALAS). These agencies have various requirements for certification and various organizational sponsors. The AALAS Technician Certification is the highest recognitions for technicians in the laboratory animal science profession. The AALAS Technician Certification program was developed to recognize professional achievement and to provide an authoritative endorsement of an technician's level of competence in the field of laboratory animal technology. The certification program is described in detail in the *Technician Certification Handbook.*

Employment Outlook

Overall, opportunity for employment as a Laboratory Animal Technician is expected to grow for some time. Technological advancement will have two opposite effects on employment: New sophisticated tests will encourage more testing and will enhance employment opportunities, but the advances in technology and laboratory automation will allow each technician to perform more tests, thus slowing down growth in this profession. The new Clinical Laboratory Improvement Act will impose increased academic standards for persons conducting certain evaluations, and job opportunities will be greater for technologists who have at least an associate degree. Fastest growth is expected in the independent laboratories, as veterinarians continue to send a large portion of their testing work to such laboratories.

Related Occupations

Occupations related to the Laboratory Technician include: Equine Veterinary Assistant and Veterinarian.

MOUNTED PARK RANGER

Description

The duty of the Mounted Park Ranger is to enforce all laws and regulations within county, state, and national parks. Mounted Park Rangers instruct, guide, and ensure the safety of all park visitors. They issue permits for vehicle use within the park, point out areas of special interest, and provide a registration program in the event a visitor becomes lost. Mounted Park Rangers also protect the cultural and natural resources and the wildlife within the park. They protect the parks from abuse by visitors by patrolling the parks on horseback to prevent vandalism, theft, fires, and danger to wildlife. They participate in search and rescue operations and the tagging and tracking of animals within the park. Finally, they are responsible for the planning and preparing of exhibits, for supervising volunteers and other workers, and for preparing and maintaining a financial budget.

Educational and Training Requirements

A high school education and a two- or four-year college degree in Land Management, Ecology, or Forestry is beneficial for this profession. While in high school, students should take courses in science, math, public speaking, and basic typing or keyboarding. College courses should include natural resource management, police science, earth science, social science, and business administration. Most federal and state Park Rangers have a two- or four-year college degree; however, it is possible to meet the requirements with a specific background of experience or with a combination of background and experience. Volunteer experience is considered work experience. Naturally, the Mounted Park Ranger must have advanced riding skills and exhibit a high degree of horsemanship.

Personal Qualifications

Mounted Park Rangers must be at least 21 years of age, have a deep interest in nature and conservation, and enjoy working outdoors. They must have good communication skills, especially when dealing with the public. They must be mature, confident, polite, tactful, and possess a calm temperament when dealing with emergencies. Mounted Park Rangers must enjoy spending long hours on horseback in all types of weather conditions. Although most work is performed outdoors, they will often be required to work indoors. They must be willing to relocate throughout their career. As most work is physically challenging, Mounted Park Rangers must be in top physical shape. They must have good eyesight, 20/100 vision or better in each eye that is correctable to 20/20 with glasses or contact lenses. Although the 40-hour week is normal for Mounted Park Rangers, they must be on call 24 hours each day, especially when the tourist population is at its highest.

Licenses and Certification

National Park Service applicants must pass the United States Federal Service entrance examination. An American Red Cross First-Aid Certificate will qualify applicants for work in water safety or lifeguard work. Candidates must possess a valid automobile driver's license and a good driving record, and they must satisfactorily complete the prescribed training required for recruits after appointment. Finally, candidates must satisfactorily complete a probationary period of one year.

Employment Outlook

Employment opportunity for the Mounted Park Ranger within the National Park Service is limited. There is strong competition for each of these openings, and as many as 100 applicants for each opening. Employment with other federal, state, and county agencies engaged in the management of land and water resources may offer more opportunities for employment.

Related Occupations

Occupations related to the Mounted Park Ranger include: Equine Extension Service Agent and Mounted Police Officer.

MOUNTED POLICE OFFICER

Description

Mounted Police Officers, like all law enforcement agents, are responsible for enforcing statutes, laws, and regulations designed to protect life and property. They spend a great deal of time in a designated area to preserve peace and to prevent crime. They may be involved in various activities such as traffic control, crowd control, patrolling parking lots, and patrolling parks and recreation areas. Mounted Police Officers are required to write reports, maintain records, and testify in court. Other duties might include the daily care and management of their equine partner and the care and maintenance of their equipment, such as bridles, saddles, halters, boots, and so on. The Mounted Police Officer is a highly visible figure in the community and has gained the respect of residents in the fight against crime. (See Figure 10.16.)

Educational and Training Requirements

The educational requirement for law enforcement has increased in recent years. A high school education is mandatory for this position. A college degree is not a requirement, but college courses in criminology, police science, criminal justice, and law enforcement may enhance

FIGURE 10.16
The duty of the Mounted Police Officer is to preserve the peace and to prevent crime.

an applicant's chances for employment. Physical education as well as sports will aid the applicant in developing the necessary physical stamina and agility to perform all skills related to law enforcement. There is usually a mandatory period of training for all recruits; in some states and large cities, recruits are required to receive training at a police academy for a 12- to 14-week period. Training includes classroom instruction on such subjects as civil rights, constitutional law, state laws, and investigative procedures, as well as training in the use of firearms, self-defense, first aid, and dealing with emergency situations. All recruits selected for the mounted units receive training in equitation skills as well as basic horsemanship skills. Finally, all applicants for the mount police units must have a keen interest in horses and the role they play in law enforcement. Naturally, any skill in horsemanship or equitation is considered an asset to the applicant seeking a position as a Mounted Police Officer.

Personal Qualifications

The Mounted Police Officer usually works a 40-hour week. Most rookie officers must work weekends, evenings, and holidays. Mounted Police Officers are required to work outdoors for long periods of time in all types of weather conditions. The profession of law enforcement may become stressful to officers as well as their families. Law enforcement officers must possess certain personal traits such as honesty, integrity, a sense of fairness, and a sense of responsibility. They should enjoy working with the public and people in general. Finally, Mounted Police Officers must have a genuine love of horses.

Licenses and Certification

All candidates for the position of police officer must adhere to civil service regulations. Candidates must be a citizen of the United States and at least 20 years of age. Eligibility for a position as a Mounted Police Officer depends on performance on a written examination as well as a physical examination. In some states, only senior officers who have served on the police force for a required amount of time are able to apply for a position with a mounted unit.

Employment Outlook

The position of a Mounted Police Officer is attractive to many people; however, the number of mounted units within police forces throughout the country is slowly decreasing due to overall budget cuts. Maintaining a stable of horse can be a costly venture for police departments on a strict budget. Because of the attractive nature of a mounted police unit, competition is very keen within police departments. Federal, state, and local law enforcement agencies have increased their standards for and become more selective in filling their mounted unit positions because the number of candidates far exceeds the number of openings. Requirements may include seniority qualifications, standard waiting lists, and recommendations from senior officers. The need to replace mounted officers who retire, transfer to other agencies, or leave a position because of a disability is the reason for most job openings within the mounted police units.

Related Occupations

Occupations related to the Mounted Police Officer include: Animal Control Officer and Mounted Park Ranger.

OUTRIDER

Description

The responsibility of the Outrider is to lead the horses to the starting gate on time. If a horse runs off or unseats a jockey while warming up, it is the duty of the Outrider to catch the runaway by riding alongside it and grabbing the reins in order to bring it under control. Once the horses enter the starting gate, the Outrider is positioned in front of the gate on the right side. If a horse breaks out of the gate prematurely, it is the duty of the Outrider to catch the horse and return it to the starting gate area for inspection by a veterinarian and reloading.

During the morning workouts, Outriders are positioned on the racetrack to observe the morning exercise sessions. It is their job to catch a loose horse or one that is out of control with the rider. By means of a two-way portable radio, Outriders can summon an ambulance and sound the alarm so that everyone is alerted to the fact that there is a loose horse or a fallen rider. The Outriders are also responsible for closing and opening the racetrack during morning training hours at designated times to allow for harrowing and training.

Educational and Training Requirements

A high school education with a good background in equitation skills as well as good horsemanship skills are essential for this position. A two-year degree in Equine Science with courses in equine behavior and anatomy would be beneficial. Any practical experience on the racetrack as a pony boy or girl or as an exercise rider is also an asset for the position of Outrider.

Personal Qualifications

The Outrider must be in excellent health and physically fit. As with other racetrack positions, Outriders must be willing to rise early and work long hours each day for 5 or 6 days per week. Most work is conducted outdoors in all types of weather conditions. Outriders must be alert at all times for any horse that is misbehaving or trying to unseat its rider. They must concern themselves with the safety of both horse and rider in each race and morning workout. Outriders must be excellent riders and possess excellent rapport with their mounts; they must trust and know the capabilities of their mounts to successfully perform their duties.

Licenses and Certification

As the Outrider is employed by the racetrack, there is no specific license or certification required for the Outrider. The only license requirements are those that apply to all racetrack employees, which entail fingerprinting, photographs, and background checks.

Employment Outlook

This position is available at racetracks offering both flat racing and harness racing. However, with the present trend of a declining number of racetracks, there will be a decline in the need for the services of an Outrider.

Related Occupations

Occupations related to the Outrider include: Exercise Rider, Jockey, and Pony Boy or Girl.

PEDIGREE RESEARCHER

Description

The duty of the Pedigree Researcher is to research the pedigrees of horses in order to determine their ancestry and their abilities. The Pedigree Researcher makes notes on individual horses within a pedigree as to their awards, earnings, show records, race records, and breeding records. In the case of breeding stallions and mares, Pedigree Researchers determine the ability of the horses to pass on their genetic traits to their offspring. They aid the Horse Breeder in determining which stallions to breed to which mares and in determining which mares to purchase to improve their breeding program. They also assist the buyer in selecting a stallion for the purpose of syndication. They may assist an appraiser or lawyer in determining the true value of a horse based on its pedigree. The Pedigree Researcher may be employed by a bloodstock agent, by a horse sales company, or act as an individual consultant.

Educational and Training Requirements

A high school education plus a vocational program emphasizing Agriculture and Livestock Production is desirable. A college degree is not a prerequisite but will definitely enhance the possibility of employment. Good Computer skills as well as Communication skills are definite requirements for this position. High school and college courses should include biology, genetics, computer science, and business. Practical experience with horses is not essential but is desirable. Any practical experience with a bloodstock agency or horse sales company will enhance the possibility of employment.

Personal Qualifications

Pedigree Researchers must be able to work independently and should be able to deal with minute details. Normal working hours are standard when employed by a bloodstock agency or horse sales company. As an independent consultant, the hours may be long and irregular when dealing with clients. Pedigree Researchers must have a pleasant personality, as well as being honest, dependable, and discreet when working with clients and fellow employees. Most work is performed indoors in an office setting, but on occasion Pedigree Researchers attend a horse sale with a client. A good deal of their time is spent working on a computer.

Licenses and Certification

There is no license or certification required for the position of Pedigree Researcher.

Employment Outlook

The employment outlook for the Pedigree Researcher is good. Pedigree research is considered a limited field, as more and more bloodstock agencies and sales companies rely on available computer data for their pedigree information. The Pedigree Researcher who becomes an expert in the field can not only compile pedigrees but also determine what individual horses in the pedigree will affect values and determine future matings for individual horses.

Related Occupations

Occupations related to the Pedigree Researcher include: Appraiser, Auctioneer, and Bloodstock Agent.

PONY BOY OR GIRL

Description

In order to understand the duties of the Pony Boy or Girl, the reader needs a basic understanding of the term *pony* and how it is applied on the racetrack. A Pony on the racetrack refers to any breed of horse used to accompany the racehorse during its morning exercise sessions. Ponies are also used to accompany the racehorse to the starting gate before a race. The person riding the pony in the mornings is referred to as a Pony Boy or Girl. The responsibility of the Pony Boy or Girl is to "pony" the racehorse during the morning exercise sessions. Ponying refers to the pony and its rider leading the riderless racehorse on the right side of the pony around the racetrack at a walk, trot, or canter. Pony Boys or Girls also accompany a fractious racehorse and its rider in the morning workouts in order to keep the racehorse under control and to keep it calm and relaxed. Before a race, the Pony Boy or Girl accompanies a racehorse and its jockey from the saddling paddock through the post parade, the warm-up period, and finally to the starting gate. A Pony Boy or Girl may be a salaried employee of the owner of several ponies; the owner, in turn, provides a ponying service to the trainers and the racing association. Some Pony Boys and Girls are independent contractors who own their own ponies. Pony Boys and Girls must provide full care to the pony or ponies that they use or own. (See Figure 10.17.)

Educational and Training Requirements

The basic educational requirement consists of a high school diploma. A good background in general horsemanship skills is required in order to

FIGURE 10.17
The Pony Boy or Girl provides exercise to the racehorse and escorts the racehorse to the starting gate.

properly care for the ponies. Any vocational training such as 4-H or FFA in the area of equine science is beneficial. The Pony Boy or Girl must have riding ability in both English and Western disciplines, and practical experience on the racetrack as a groom, exercise rider, and so on is very helpful.

Personal Qualifications

Pony Boys and Girls must be physically fit and in good health. They must be willing to rise early and work long hours each day for a 6- or 7-day workweek. They must be willing to care for their mounts on a daily basis, which includes exercising, tacking, cleaning the stall, feeding, watering, bathing, and administering medication. They must also be willing to travel to various racetracks throughout the year. Most of their work is performed outdoors in all types of weather conditions.

Licenses and Certification

As with any racetrack employee, Pony Boys and Girls must be licensed by the racetrack or the state racing commission at the location where they are employed. Licensing at the racetrack

entails a background check, photographs, fingerprints, and payment of the required fee.

Employment Outlook

Horse racing is presently suffering from a decline in the number of racetracks as well as the number of horses in competition at each track. This results in a decline in the need for a Pony Boy or Girl. In addition to racetracks, however, training centers and breeding farms may offer employment opportunities for the skilled Pony Boy or Girl.

Related Occupations

Occupations related to the Pony Boy or Girl include: Exercise Rider, Groom, Hot Walker, Jockey, and Outrider.

RETAIL TACK SHOP OPERATOR

Description

The responsibility of the Retail Tack Shop Operator is to market horse-related products to the horse owner and the general public. These products may include riding apparel and equipment, grooming equipment and tools, horse health products, and general stable equipment. Some Retail Tack Shop Operators specialize in one discipline, such as Western equitation, racing, or dressage; they must be knowledgeable in the discipline in which their shop specializes. Retail Tack Shop Operators must supervise all employees in pricing goods, setting up displays, and organizing shelves; inspect merchandize as to its quality; and keep a complete inventory of all goods in their shop. They must be able to assist customers in decisions concerning their horses. They are also responsible for setting and implementing all policies, goals, and procedures concerning the overall operation of the tack shop.

Educational and Training Requirements

The basic educational requirement consists of a high school diploma. A two- or four-year college degree in Equine Science or Business Management is beneficial. College-level courses should include marketing, economics, accounting, and computer science. Any practical experience in a retail trade position such as a sales clerk, store manager, or customer service representative is most helpful.

Personal Qualifications

The Retail Tack Shop Operator must have a good business sense, strong public relations skills, and customer service skills. Most work is performed indoors within the store. Hours are regular throughout the year, but irregular hours are the norm during seasonal sales and holidays. Work is usually performed during the day as well as evenings and weekends. Retail Tack Shop Operators can set their own hours, but all hours must be convenient for the customers.

Licenses and Certification

There is no mandatory license or certification requirement for the Retail Tack Shop Operator.

Employment Outlook

The employment outlook for the Retail Tack Shop Operator is for the most part poor, as more and more horse owners are making their purchases from discount catalog retailers. Competition with other tack shops in the same area may lead to bankruptcy. However, the location of the retail tack shop can make the difference between success and failure. As with any retail business, the condition of the overall economy also plays a key role in determining the success of the business.

Related Occupations

Occupations related to the Retail Tack Shop Operator include: Equine Book Dealer, Farm Equipment Dealer, Feed and Grain Dealer, and Saddlesmith.

RIDING INSTRUCTOR

Description

The responsibility of the Riding Instructor is to teach equitation on various levels from novice to advanced. Riding Instructors may become specialized in one discipline of riding, such as dressage, hunters and jumpers, English equitation, saddle seat, and Western pleasure. In addition to teaching riding skills, good Riding Instructors instruct their students in areas such as safety, riding equipment, grooming and tacking procedures, and basic stable management skills.

Riding Instructors may be employed by a farm or stable, or they may work as independent instructors. If they are employed by a farm or stable, they usually receive a fee or a commission for each riding lesson given. Some receive a small salary as well as living quarters. The independent Riding Instructor may work out of a leased or self-owned facility. In this situation, they either supply the school horses or allow students to transport their own horses to the Instructor's facility. Independent Riding Instructors may also opt to travel to a facility owned and operated by their students to provide private lessons with the students' horses. (See Figure 10.18.)

Educational and Training Requirements

A high school education as well as formal training in equitation and stable management are

FIGURE 10.18
The Riding Instructor teaches equitation on various levels, from novice to advanced.

helpful for this profession. A two- or four-year college degree in Equine Science or Animal Science is also beneficial. Courses in education and business management will aid the applicant in a position as a Riding Instructor.

Personal Qualifications

Riding Instructors must be willing to rise early and work long hours each day. It is not unusual for the Riding Instructor to work a 7-day week. Riding Instructors must be willing to travel and act as a coach to their students at various competitions. They may be required to work indoors as well as outdoors in all types of weather conditions. Riding Instructors must have good communication skills and the ability to motivate their students. They must be in good physical condition and in excellent health. They also must be organized and have a pleasant and patient attitude when dealing with students of various temperaments and age groups.

Licenses and Certification

There is no specific licensing requirement for Riding Instructors, although they may become certified through various riding organizations. Some organizations that offer certification are the Certified Horsemanship Association (CHA), the North American Riding for the Handicapped Association (NARHA), the United States Dressage Federation (USDF), and the American Riding Instructors Association (ARIA). According to the ARIA, their certification process is accomplished by evaluating Riding Instructors' qualifications and teaching ability through written and oral testing and a videotaped lesson. Instructors who pass these tests and who teach in a safe and competent manner earn certification. The American Riding Instructors Certification Program (ARICP) offers certification to instructors in three levels of experience and in eleven teaching specialties. To ensure that the ARICP Certified Instructors standards remain high, recertification is required every five years. This also gives Instructors an opportunity to upgrade their level of certification and to add new teaching specialties.

Employment Outlook

The employment outlook for the Riding Instructor is promising. Each year more people in the United States are turning to horseback riding lessons as a form of exercise and recreation. This will result in an increased demand for qualified Riding Instructors. The qualified professional Riding Instructor must compete with the inexperienced novice instructor who will provide riding lessons for a modest fee. These inexperienced Riding Instructors are also willing to give riding lessons to students in exchange for free board for their horse at a farm or stable. The average, uneducated horse-loving public is usually ill-equipped to determine the ability of their Riding Instructor. However, good experienced Riding Instructors will usually do well.

Related Occupations

Occupations related to the Riding Instructor include Equine Educator, Farm or Ranch Manager, and Horse Trainer.

SADDLESMITH

Description

The duty of the Saddlesmith is to create leather horse products such as saddles, bridles, and harnesses. Saddlesmiths check the texture, strength, and color of leather; apply leather dyes and liquid topcoats to produce a high-gloss finish to saddles, bridles, and harnesses; and decorate the surface of leather products by hand stitching or by stamping the leather with decorative patterns and designs. Saddlesmiths also repair leather by using hand tools and machines designed for sewing, hole punching, and nailing. They may be employed by a custom tack manufacturer, or a tack shop, or they may be self-employed. Most work is performed indoors with little, if any, exposure to horses.

Educational and Training Requirements

A high school education is a requirement for this position. A Saddlesmith usually learns the trade either through an in-house training program or by working as an apprentice to a seasoned, established saddle maker. There are several vocational schools that offer training in leather work and repair. These vocational programs specialize in basic skills such as cutting, stitching, and dying leather in addition to saddle, bridle, and harness making. Any practical experience with horses is definitely an asset to this profession.

Personal Qualifications

The Saddlesmith must have good manual dexterity as well as the mechanical ability to work with hand tools and machines. Custom Saddlesmiths must also have a certain amount of artistic ability. They must have good communication skills and a pleasant personality when dealing with clients and fellow workers. They must be able to work independently with little supervision. Saddlesmith employed by custom leather manufacturers generally work a normal 40-hour workweek. Those who are self-employed or work in tack shops work irregular hours, including weekends and evenings.

Licenses and Certification

No mandatory license or certification is required for the profession of a Saddlesmith.

Employment Outlook

The employment outlook for the Saddlesmith appears to be poor because of the introduction of inexpensive imports of saddles, bridles, and harnesses to the United States. This profession is also affected by the rising costs of leather products. Finally, leather equestrian goods are being replaced by cheaper substitutes such as vinyl and nylon. However, some of the more expensive, fine leather equestrian products will continue to be repaired, which will result in the employment of the Saddlesmith in the capacity of a leather repairer.

Related Occupations

Occupations related to the Saddlesmith include: Retail Tack Shop Operator.

STARTER AND ASSISTANT STARTER

Description

The Starter is responsible for a fair and safe start for every horse race. At a flat racing meet, an electrical starting gate is used to officially start a horse race. At a harness horse racing meet, a gate mounted on an automobile is used to start a race for trotters and pacers. At a steeplechase meet, a lining-up method is used to start a race. The Assistant Starter works directly with the Starter and is responsible for actually loading the horse into the starting gate in flat races. In addition to starting races, the Starter and Assistant Starter must be available in the early morning hours on the racetrack to assist trainers in gate schooling sessions for young inexperienced horses. The Starter must also keep accurate records pertaining to the starting gate for each horse stabled on the racetrack.

Educational and Training Requirements

A high school education with some vocational training in general horsemanship is essential for both the Starter and Assistant Starter. A two- or four-year degree in Equine Science or Animal Science is beneficial but is not a requirement. A knowledge of equine behavior and equine anatomy and physiology is also helpful for this profession. The Starter should have a thorough understanding of the horse racing industry and should also serve an internship period as an Assistant Starter at a racetrack to gain valuable practical experience. The Assistant Starter must have a basic knowledge of equine behavior as well as good horsemanship skills.

Personal Qualifications

Both Starters and Assistant Starters must be physically fit and in excellent health in order to handle horses on a daily basis, and they must be able to make quick decisions concerning the safety of the jockeys, drivers, and horses. They must have good communication skills in order to relate to trainers, jockeys, stewards, and racing officials. They also must have keen eyesight as well as good judgment and common sense. Above all, Starters and Assistant Starters must have patience when working with nervous horses. They must be willing to work long hours and work in all types of weather conditions.

Licenses and Certification

There is no formal licensing or certification requirement for the Starter and Assistant Starter. However, both of these professions require a strong background in the horse racing industry. An Assistant Starter may be employed as a groom, hot walker, exercise rider, or assistant trainer before becoming part of the starting gate crew. The Starter also needs to have a diverse background on the racetrack in addition to having experience as an Assistant Starter before being hired as a Starter by a racetrack.

Employment Outlook

The employment outlook for the Starter and Assistant Starter, like other occupations in the horse racing industry, is poor. However, with new racetracks being built, employment opportunities for these professions may be on the rise in the near future.

Related Occupations

Occupations related to the Starter and Assistant Starter include Exercise Rider, Groom, Horse Trainer, and Hot Walker.

VETERINARIAN

Description

Veterinarians specializing in horses are responsible for diagnosing medical problems, treating wounds, mending broken bones, performing surgery, prescribing and administering medications, and vaccinating horses against diseases. They also advise horse owners on the care, feeding, breeding, and management of horses. Equine Veterinarians may become specialists in areas such as racing, breeding, nutrition, pharmaceuticals, research, and education. Equine Veterinarians contribute to the health of humans as well as horses. They prevent the outbreak and spread of equine diseases, such as rabies, that can be transmitted to humans. They are also required to perform autopsies on deceased horses to determine the cause of death. Most Veterinarians who specialize in horses are involved in a private practice. Some Veterinarians are engaged in a general practice and treat other types of animals in addition to horses. (See Figure 10.19.)

Educational and Training Requirements

Students must complete at least six years of college in order to receive a doctorate of

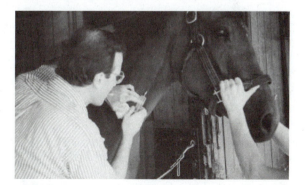

FIGURE 10.19
The Veterinarian is responsible for maintaining the general health of the horse.

Veterinary Medicine (DVM) degree, which includes two years of preveterinary training with an emphasis on the biological and physical sciences and a four-year degree in Veterinary Medicine from an accredited veterinary college. Most applicants to veterinary programs have already completed four years of college. Course requirements prior to entry into veterinary school include biology, zoology, chemistry, biochemistry, microbiology, language arts, humanities, and social sciences. Course requirements for the first two years include biochemistry, anatomy, physiology, microbiology, pathology, pharmacology, as well as other veterinary-related courses. The last two years in Veterinary Medicine include courses such as surgery, ethics and medical law, obstetrics, reproductive diseases, radiation biology, applied and clinical medication, and public health.

Entrance into schools of Veterinary Medicine is highly competitive because there are only 27 accredited colleges of Veterinary Medicine in the United States. These colleges are accredited by the Council of Education of the American Veterinary Medical Association (AVMA). Applicants for admission must take the Veterinary Aptitude, Medical College Admission Test or the Graduate Record Examination. Applicants must also submit written evidence that they have had practical experience with animals.

Personal Qualifications

People who wish to become an equine Veterinarian must have a genuine love for both horses and people. Veterinarians must have excellent communications skills, both written and oral, when dealing with horse owners and other professional horse people. They must have good business sense, especially those who plan to develop a private practice. Veterinarians specializing in horses usually work out of a fully equipped mobile clinic. They are required to drive long distances, especially in rural areas, and they are required to work long hours (average of 50 per week, including evenings, holidays, and weekends) with little or no time off.

Equine practitioners are always in danger of being kicked, bitten, and stepped on by their equine patients. They may also be exposed to contagious diseases as well as general infections and hazardous substances. Most of their work is performed outdoors in various types of climate and weather conditions.

Licenses and Certification

All Veterinarians must be licensed in order to practice veterinary medicine. Licensing requires an applicant to possess a DVM or VMD degree from an accredited college. Applicants must successfully pass both an oral and a written examination, issued by both state and national agencies. Usually, a state education department, agricultural department, or other state agency will have the authority to issue licenses. Some states issue a license to Veterinarians already licensed in another state. For the most part, Veterinarians do not have to serve as interns, as do physicians who treat humans, before they begin to practice veterinary medicine; however, some states do require an internship period before a license can be issued. Most states require Veterinarians to continue their education by attending seminars and courses on the latest medical technology and treatments. Veterinary specialists, such as surgeons and cardiologists may become board certified through the American Veterinary Medical Association or through their affiliated state or local chapters.

Employment Outlook

The employment outlook for the equine Veterinarian is good. Most horse owners and breeders have high regard for their horses' health. They are diligent about making sure their horses have received required vaccinations and wish to provide the best health care possible. These people will always require the services of a professional Veterinarian for their horses. Prospects for Veterinarians who specialize in horses in rural areas are better than average, since most

Veterinarians prefer working in large metropolitan areas. The growing emphasis on scientific horse breeding and horse management techniques will undoubtedly contribute to the demand for the professional equine Veterinarian. In addition to a private equine practice, Veterinarians may find employment opportunities in areas such as government, public health, military, industry research, teaching, and zoology and wildlife. Positions for newly licensed Veterinarians will open as Veterinarians retire, die, or leave the profession.

Related Occupations

Occupations related to the Veterinarian include Breeding Technician and Equine Geneticist, Equine Veterinary Assistant, and Veterinary Pharmaceutical Salesperson.

VETERINARY PHARMACEUTICAL SALESPERSON

Description

The Veterinary Pharmaceutical Salesperson is employed by a veterinary pharmaceutical company to assist veterinarians in the selection and purchase of pharmaceuticals. The primary function of this Salesperson is to assist veterinarians in determining what type of drugs they will need for their equine practice. Veterinary Pharmaceutical Salespersons may describe features of drugs, deliver purchases, and prepare pharmaceutical displays. They must be aware of the promotions their company is sponsoring, and also of those being sponsored by their competitors. They are usually assigned a territory and, in some cases, are provided with leads from the pharmaceutical company they represent. Veterinary Pharmaceutical Salespersons may receive a basic salary plus a commission on their overall sales. Some pharmaceutical companies provide their sales force with a company vehicle to service their veterinarian clients.

Educational and Training Requirements

For this profession a high school education is required, along with a two- or four-year degree in preveterinary studies. Equine Science or Biochemistry is recommended. College-level courses should include toxicology, chemistry, marketing, and business management. Some veterinary pharmaceutical companies train their sales force about their products and services.

Personal Qualifications

Veterinary Pharmaceutical Salespersons usually transact all business indoors, but on occasion they meet with veterinarians outdoors. Their hours are long and irregular, as they must meet with veterinarians when they are available from their practice. Salespersons must be willing to drive and travel long distances each day in order to cover their assigned territory, and they must be willing to work at least 6 days per week. Veterinary Pharmaceutical Salespersons attend veterinary seminars and conventions in order to keep abreast of new concepts and treatments in veterinary medicine. They also attend horse-related expositions in order to promote their products to veterinarians and the general public. They must have a pleasant personality when dealing with both clients and superiors. Among other desirable characteristics is an interest in sales work, a clean and neat appearance, and the ability to communicate clearly and effectively.

Licenses and Certification

There is no special licensing or certification requirement for this profession. Those Salespersons serving veterinarians located on racetracks must be licensed by the racetrack in order to gain access to the stable area. A state driver's license is mandatory.

Employment Outlook

The employment outlook for the Veterinary Pharmaceutical Salesperson is good. The veterinary pharmaceutical industry is very lucrative and is presently experiencing phenomenal growth. Companies specializing in human pharmaceuticals are now creating veterinary divisions in order to meet the needs of the expanding veterinary medicine market. This type of growth will require a large sales force in the near future. In some geographic areas of the country, companies are faced with a shortage of qualified applicants for sales positions. As a result, companies can be expected to improve efforts to attract and retain Salespersons by offering higher wages, generous benefits, higher sales commissions, and more flexible schedules.

Related Occupations

Occupations related to the Veterinary Pharmaceutical Salesperson include: Equine Veterinary Assistant, Laboratory Animal Technician, and Veterinarian.

Appendix

Directory of Professional Equine Resources

ANIMAL CONTROL OFFICER

American Horse Protection Association, Inc.
1000 29th Street NW, T-100
Washington, DC 20037-3820

American Humane Society
63 Inverness Drive East
Englewood, Colorado 80112
http://www.americanhumane.org

Hooved Animal Society
PO Box 400
Woodstock, Illinois 60098
http://www.hahs.org

ANNOUNCER

National Association of
Broadcasters
1771 North Street NW
Washington, DC 20036
http://www.nab.org

APPRAISER

American Society of Equine
Appraisers
PO Box 186
Twin Falls, Idaho 83303
http://www.equineappraiser.com

ARCHITECT

American Institute of Architects
1735 New York Avenue NW
Washington, DC 20006
http://www.aiaonline.com

ASSISTANT HORSE TRAINER

University of Arizona
Race Track Industry Program
Educational Building #69,
Room 104
Tucson, Arizona 85721-0069
http://www.arizona.edu

AUCTIONEER

Fasig Tipton Sales Company
2400 Newtown Pike
Lexington, Kentucky 40583
http://www.fasigtipton.com

Keeneland Sales
PO Box 1690
Lexington, Kentucky 40588
http://www.keeneland.com

National Auctioneers Association
8880 Ballentine
Overland Park, Kansas 66214
http://www.auctioneers.org

World Wide College of Auctioneering
PO Box 949, Department WHD
Mason City, Iowa 50402
http://www.wwcollege.qpg.com

AUCTION SALES EMPLOYEE

Fasig Tipton Sales Company
2400 Newtown Pike
Lexington, Kentucky 40583
http://www.fasigtipton.com

Keeneland Sales
PO Box 1690
Lexington, Kentucky 40588
http://www.keeneland.com

National Equine Sales, Inc.
2040 West Blee Road
Springfield, Ohio 45502
http://www.nationalequine.com

BLOODSTOCK AGENT

Livestock Marketing Association
7509 Tiffany Springs Parkway
Kansas City, Missouri 64153
http://www.lmaweb.com

National Equine Sales, Inc.
2040 West Blee Road
Springfield, Ohio 45502
http://www.nationalequine.com

BREEDING TECHNICIAN

Genetics Society of America
9650 Rockville Pike
Bethesda, Maryland 20814
http://www.intl.genetics.org

National Association of Animal Breeders
PO Box 1033
Columbia, Missouri 65205
http://www.naab-css.org

CLOCKER

National Thoroughbred Racing Association
2343 Alexandria Drive #210
Lexington, Kentucky 40504
http://www.ntraracing.com

EQUINE ACCOUNTANT OR BOOKKEEPER

American Institute of Certified Public
Accountants
1211 Avenue of Americas
New York, New York 10036
http://www.aicpa.org

National Society of Public Accountants
1010 North Fairfax Street
Alexandria, Virginia 22314
http://www.nsacct.org

EQUINE ADVERTISING SPECIALIST

American Marketing Association
250 South Wacker Drive, Suite 200
Chicago, Illinois 60606
http://www.ama.org

EQUINE ARTIST

American Academy of Equine Artists
PO Box 1315
Middleburg, Virginia 22117
http://www.aaea.net

EQUINE ATTORNEY

Horse Law News
PO Box 579
Redwood City, California 94064
http://www.piebaldpress.com

Law School Admission Test
PO Box 2000
Newton, Pennsylvania 28940
http://www.lsac.org

EQUINE BOOK DEALER

American Booksellers Association
282 South Broadway
Tarrytown, New York 10591
http://www.bookweb.org

Directory of American Booksellers
560 White Plains Road
Tarrytown, New York 10591

EQUINE CHIROPRACTOR

American Veterinary Chiropractic Association
623 Main Street
Hillsdale, Illinois 61257
http://www.animalchiropractic.org

EQUINE CONSULTANT

North American Horseman's Association
310 Wahburne Avenue
Paynesville, Minnesota 56262
http://www.arkagency-naha.com

Professional Horseman's Association of America
2009 Harris Road
Penfield, New York 14526

EQUINE DENTIST

American Veterinary Medical Association
1931 North Meacham Road #100
Schaumburg, Illinois 60173
http://www.avma.org

Association for Equine Dental Equilibration
PO Box 970
Glenns Ferry, Idaho 83623

EQUINE EDUCATOR

American Association of Colleges for Teacher Education
1 Dupont Circle, Suite 610
Washington, DC 20036
http://www.aacte.org

American Association of University Professors
1012 14th Street, NW, Suite 500
Washington, DC 20005
http://www.aaup.org

American Vocational Association
1410 King Street
Alexandria, Virginia 22314

EQUINE EXTENSION SERVICE AGENT

National Association of Extension 4-H Agents
University of Georgia
College of Agriculture
Cooperative Extension Service
Athens, GA 30602
http://www.uga.edu

United States Department of Agriculture
Extension Service
14th and Independence Avenue SW
Washington, DC 20250
http://www.usda.gov

EQUINE FENCE DEALER OR INSTALLER

American Fence Association
2336 Wisteria Drive, Suite 230
Snellville, Georgia 30078
http://www.americanfenceassociation.com

EQUINE GENETICIST

Genetics Society of America
9650 Rockville Pike
Bethesda, Maryland 20814
http://intl.genetics.org

National Association of Animal Breeders
PO Box 1033
Columbia, Missouri 65205
http://www.naab-css.org

EQUINE JOURNALIST

American Horse Publications
49 Spinnaker Circle
South Daytona, Florida 32119
http://www.americanhorsepubs.org

National Turf Writers Association
1314 Bentwood Way
Louisville, Kentucky 40223

United States Harness
Writers Association, Inc.
PO Box 10
Batavia, New York 14021
http://www.ushwa.trot.net

EQUINE LIBRARIAN

American Association of Museums
1225 Eye Street NW, Suite 200
Washington, DC 20005
http://www.aam-us.org

EQUINE LOAN OFFICER

American Bankers Association
1120 Connecticut Avenue NW
Washington, DC 20036
http://www.aba.com

American Farm Bureau Federation
225 Touhy Avenue
Park Ridge, Illinois 60068
http://www.fb.com

EQUINE MASSAGE THERAPIST

Equine Massage Therapy Information Center
7074 Ducketts Lane
Elkridge, Maryland 21075

International Association of Equine Sport
Massage Therapists
PO Box 447
Round Hill, Virginia 22141
http://www.iaesmt.com

International Federation of Registered Equine
Massage Therapists
PO Box 46019
956 Dundas Street
London, Ontario, Canada NSW 3A1

EQUINE NUTRITIONIST

American Feed Industry Association
1501 Wilson Boulevard #100
Arlington, Virginia 22209
http://www.afia.org

Equine Nutrition and Physiology Society
309 West Clark Street
Champaign, Illinois 61820
http://www.enps.org

EQUINE PHOTOGRAPHER

Photographer's Market
Writer's Digest Books
1507 Dana Avenue
Cincinnati, Ohio 45207
http://www.writersdigest.com

Professional Photographers of America, Inc.
57 Forsythe Street #1600
Atlanta, Georgia 30303
http://www.ppa.com

EQUINE REAL ESTATE BROKER AND AGENT

American Association of Certified Real Estate
Appraisers
800 Compton Road, Suite 10
Cincinnati, Ohio 45231

American Real Estate Society
Cleveland State University
Department of Finance
Cleveland, Ohio 44115
http://www.aresnet.org

National Association of Real Estate Brokers
1629 K Street, Suite 605
Washington, DC 20006
http://www.nareb.com

EQUINE RECREATION DIRECTOR

National Association of State Park Directors
126 Mill Branch Road
Tallahassee, Florida 32312
http://www.indiana.edu/~naspd/

National Recreation and Park Association
3101 Park Center Drive
Alexandria, Virginia 22302
http://www.nrpa.org

EQUINE SPORTSCASTER

National Association of Broadcasters
1771 North Street, NW
Washington, DC 20036
http://www.nab.org

Radio-Television News Directors Association
1717 K Street NW, Suite 615
Washington, DC 20006
http://www.rtnda.com

EQUINE TRANSPORTATION SPECIALIST

American Association of Exporters and
Importers
11 West 42nd Street, 13th Floor
New York, New York 10036
http://www.americanimporters.org

Animal Transportation Association
10700 Richmond, Suite 201
Houston, Texas 77042
http://www.npscmgmt.com/AATA/

National Horse Carriers Association
2053 Buck Lane
Lexington, KY 40511
http://www.nationalhorsecarriers.com

EQUINE TRAVEL AGENT

American Society of Travel Agents
Education Department
1101 King Street
Alexandria, Virginia 22314
http://www.astanet.com

Institute of Certified Travel Agents
148 Linden Street
Wellesley, Massachusetts 02181
http://www.icta.com

EQUINE VETERINARY ASSISTANT

American Animal Hospital Association
60 Garland Drive
Northglenn, Colorado 80233
http://www.aahanet.org

American Association of Equine Practitioners
4075 Iron Works Parkway
Lexington, Kentucky 40511
http://www.aaep.org

American Veterinary Medical Association
1931 North Meacham Road #100
Schaumburg, Illinois 60173
http://www.avma.org

North American Veterinary Technician
Association
PO Box 224
Battleground, Indiana 47920
http://www.avma.org/navta/

EQUINE VIDEOGRAPHER

Equine Education Institute
PO Box 68
Ringwood, Illinois 60072

Federal Communications Commission
Consumer Assistance Office
1270 Fairfield Road
Gettysburg, Pennsylvania 17325
http://www.fcc.gov

EXERCISE RIDER

National Thoroughbred Racing Association
230 Lexington Green Circle #310
Lexington, Kentucky 40503
http://www.ntraracing.com

FARM EQUIPMENT DEALER

National Association of Trailer Manufacturers
LaCosta Green Office Building
1033 LaPosad Drive, Suite 200
Austin, Texas 78752
http://www.natm.com

North American Equipment Dealers Association
10877 Watson Road
St. Louis, Missouri 63127
http://www.naeda.com

FARM OR RANCH MANAGER

American Farm Bureau Federation
225 Touhy Avenue
Park Ridge, Illinois 60068
http://www.fb.com

United States Department of Agriculture
Director of Public Information
PO Box 2890
Washington, DC 20013

FARRIER

American Farrier's Association
4089 Iron Works Pike
Lexington, Kentucky 40511

Brotherhood of Working Farriers Association
14013 East Hwy 136
LaFayette, Georgia 30728

World Farrier's Association
PO Box 1102
Albuquerque, New Mexico 87103
http://www.amfarriers.com

FEED AND GRAIN DEALER

American Feed Industry Association
1501 Wilson Boulevard #100
Arlington, Virginia 22209
http://www.afia.org

National Hay Association, Inc.
102 Treasure Island Causeway
St. Petersburg, Florida 33706
http://www.nhahay.org

FOALING ATTENDANT

North American Horseman's Association
310 Wahburne Avenue
Paynesville, Minnesota 56262
http://www.arkagency-naha.com

Thoroughbred Owners and Breeders Association
PO Box 4367
Lexington, Kentucky 40544
http://www.toba.org

GROOM

Professional Horsemen's Association of America
2009 Harris Road
Penfield, New York 14526

United Thoroughbred Trainers of America, Inc.
PO Box 7065
Louisville, Kentucky 40257
http://www.thebackstretch.com

GROUNDSKEEPER

The Professional Grounds Management Society
120 Cockeysville Road, Suite 104
Hunt Valley, Maryland 21031

HARNESS HORSE DRIVER

United States Trotting Association
750 Michigan Avenue
Columbus, Ohio 43215
http://www.ustrotting.com

HIPPOTHERAPIST

American Hippotherapy Association
5001 Woodside Road
Woodside, California 94062

North American Riding for the Handicap
Association
PO Box 33150
Denver, Colorado 80223
http://www.narha.org

HORSE BREEDER

American Quarter Horse Association
PO Box 200
Amarillo, Texas 79168
http://www.americanquarterhorse.org

National Association of Animal Breeders
PO Box 1033
Columbia, Missouri 65205
http://www.naab-css.org

Thoroughbred Owners and Breeders Association
PO Box 4367
Lexington, Kentucky 40544
http://www.toba.org

HORSE INDUSTRY SECRETARY

American Farm Bureau Federation
600 Maryland Avenue, SW, Suite 800
Washington, DC 20024

Horse Industry Alliance
8314 White Settlement Road
Fort Worth, Texas 76108

HORSE INSURANCE AGENT OR BROKER

Insurance Information Institute
110 William Street
New York, New York 10038
http://www.iii.org

HORSE INSURANCE CLAIMS ADJUSTER

Insurance Information Institute
110 William Street
New York, New York 10038
http://www.iii.org

National Association of Public Adjustors
300 Water Street
Baltimore, Maryland 21202

HORSE INSURANCE UNDERWRITER

Insurance Information Institute
110 William Street
New York, New York 10038
http://www.iii.org.

HORSE SHOW AND EVENT JUDGE

American Horse Show Association, Inc.
4047 Iron Works Parkway
Lexington, Kentucky 40511
http://www.ahsa.org

Intercollegiate Horse Show Association
PO Box 741
Stonybrook, New York 11790

HORSE TRAILER AND VAN DEALER

Animal Transportation Association
10700 Richmond, Suite 201
Houston, Texas 77042
http://www.npscmgmt.com/AATA/

National Association of Trailer Manufacturers
LaCosta Green Office Building
1033 LaPosad Drive, Suite 200
Austin, Texas 78752

North American Equipment Dealers Association
10877 Watson Road
St. Louis, Missouri 63127

HORSE TRAINER

Professional Horseman's Association of America
2009 Harris Road
Penfield, New York 14526

United States Trotting Association
750 Michigan Avenue
Columbus, Ohio 43215
http://www.ustrotting.com

United Thoroughbred Trainers of America
PO Box 7065
Louisville, Kentucky 40257

HORSE VAN OR TRUCK DRIVER

American Trucking Associations, Inc.
2200 Mill Road
Alexandria, Virginia 22314

Professional Truck Driver Institute of America
8788 Elk Grove Boulevard, Suite #20
Elk Grove, California 95624

HOT WALKER

National Horsemen's Benevolent and Protection Association, Inc.
2875 NE 191 Street, Suite 506
Aventura, Florida 33180

United Thoroughbred Trainers of American
PO Box 7065
Louisville, Kentucky 40257

IDENTIFIER

National Thoroughbred Racing Association
230 Lexington Green Circle #310
Lexington, Kentucky 40503
http://www.ntraracing.com

United States Trotting Association
750 Michigan Avenue
Columbus, Ohio 43215
http://www.ustrotting.com

University of Arizona
Race Track Industry Program
Educational Building #69, Room 104
Tucson, Arizona 85721-0069

JOCKEY

Jockeys' Guild, Inc.
PO Box 250
Lexington, Kentucky 40588
http://www.jockeysguild.com

JOCKEY AGENT

Jockeys' Guild, Inc.
PO Box 250
Lexington, Kentucky 40588
http://www.jockeysguild.com

JOCKEY VALET

Jockeys' Guild, Inc.
PO Box 250
Lexington, Kentucky 40588
http://www.jockeysguild.com

LABORATORY ANIMAL TECHNICIAN

American Association for Accreditation of
Laboratory Animal Care
11300 Rockville Pike, Suite 1211
Rockville, Maryland 20852
http://www.aaalac.org

American Association for Laboratory Animal
Science
9190 Crestwyn Hills Drive
Memphis, Tennessee 38125
http://www.aalas.org

Institute for Laboratory Animal Research
2101 Constitution Avenue
Washington, DC 20048

MOUNTED PARK RANGER

National Association of State Park Directors
126 Mill Branch Road
Tallahassee, Florida 32312
http://www.indiana.edu/~naspd

National Recreation and Park Association
3101 Park Center Drive
Alexandria, Virginia 22302
http://www.nrpa.org

United States Department of Interior
18th and C Streets NW
Washington, DC 20240
http://www.doi.gov

MOUNTED POLICE OFFICER

United States Park Police
Department of Interior
1100 Ohio Drive, SW
Washington, DC 20024
http://www.doi.gov/usparkpolice/

OUTRIDER

National Thoroughbred Racing Association
230 Lexington Green Circle, #310
Lexington, Kentucky 40503
http://www.ntraracing.com

United States Trotting Association
750 Michigan Avenue
Columbus, Ohio 43215
http://www.ustrotting.com

PEDIGREE RESEARCHER

Fasig Tipton Sales Company
2400 Newtown Pike
Lexington, Kentucky 40583
http://www.fasigtipton.com

Jockey Club
821 Corporate Drive
Lexington, Kentucky 40503
http://www.jockeyclub.com

Keeneland Sales
PO Box 1690
Lexington, Kentucky 40588
http://www.keeneland.com

PONY BOY OR GIRL

American Quarter Horse Race Track Association
PO Box 200
Amarillo, Texas 79168

National Horsemen's Benevolent and Protection
Association, Inc.
2875 NE 191 Street, Suite 506
Aventura, Florida 33180

Thoroughbred Horsemen's Association, Inc.
10500 Little Patuxent Parkway, #420
Columbia, Maryland 21044

RETAIL TACK SHOP OPERATOR

Saddle, Harness and Allied Trade Association
191 Lyman Street, Suite 1
Asheville, North Carolina 28801

Western-English Retailers Association
PO Box 430
Denver, Colorado 80216
http://www.wera.org

RIDING INSTRUCTOR

American Association of Riding Schools
8375 Coldwater Road
Davison, Michigan 48423
http://www.ucanride.com

American Riding Instructors Association
28801 Trenton Court
Bonita Springs, Florida 34134
http://www.riding-instructor.com

SADDLESMITH

Harness Shop News
Route 1, Box 212
Sylva, North Carolina 28779

Saddle, Harness and Allied Trades Association
191 Lyman Street, Suite 1
Asheville, North Carolina 28801

STARTER AND ASSISTANT STARTER

National Thoroughbred Racing Association
230 Lexington Green Circle #310
Lexington, Kentucky 40503
http://www.ntraracing.com

United States Trotting Association
750 Michigan Avenue
Columbus, Ohio 43215
http://www.ustrotting.com

VETERINARIAN

American Association of Equine Practitioners
4075 Ironworks Pike
Lexington, Kentucky 40511
http://www.aaep.com

American Veterinary Medical Association
1931 North Meacham Road #100
Schaumburg, Illinois 60173
http://www.avma.org

VETERINARY PHARMACEUTICAL SALESPERSON

American Association of Colleges of Pharmacy
1426 Prince Street
Alexandria, Virginia 22314
http://www.aacp.org

Food and Drug Administration
Center of Veterinary Medicine
Rockville, Maryland 20857
http://www.fda.gov

Index